Best Practices for Datacom Facility Energy Efficiency

Second Edition

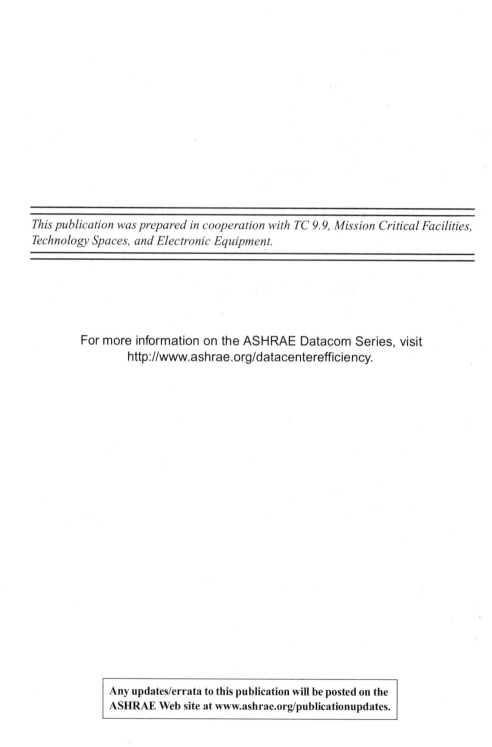

This publication was prepared in cooperation with TC 9.9, Mission Critical Facilities, Technology Spaces, and Electronic Equipment.

For more information on the ASHRAE Datacom Series, visit
http://www.ashrae.org/datacenterefficiency.

Best Practices for Datacom Facility Energy Efficiency

Second Edition

American Society of Heating, Refrigerating and Air-Conditioning Engineers, Inc.

ISBN 978-1-933742-47-2

All rights reserved. First edition 2008
Second edition 2009

Printed in the United States of America

Printed on 30% post-consumer waste using soy-based inks.

Cover photo courtesy of http://philip.greenspun.com.

Library of Congress Cataloging-in-Publication Data

Best practices for datacom facility energy efficiency.—2nd ed.
    p. cm.
  Includes bibliographical references and index.
  Summary: "The intent of this publication is to provide the reader with detailed information on the design of datacom facilities that will aid in minimizing the life-cycle cost to the client and to maximize energy efficiency in a facility to align with ASHRAE's stated direction to lead the advancement of sustainable building design and operations"—Provided by publisher.
  ISBN 978-1-933742-47-2 (softcover)
  1. Data processing service centers—Design and construction. 2. Electronic data processing departments--Equipment and supplies—Protection. 3. Buildings—Environmental engineering. 4. Data processing service centers--Energy conservation. 5. Data processing service centers—Cost of operation. I. American Society of Heating, Refrigerating and Air-Conditioning Engineers.

  TH6057.I53B47 2009
  725'.23—dc22
                            2008051512

ASHRAE Staff

Special Publications

Mark Owen
Editor/Group Manager

Cindy Sheffield Michaels
Managing Editor

James Madison Walker
Associate Editor

Amelia Sanders
Assistant Editor

Elisabeth Parrish
Assistant Editor

Michshell Phillips
Editorial Coordinator

Publishing Services

David Soltis
Group Manager

Tracy Becker
Graphic Applications Specialist

Jayne Jackson
Publication Traffic Administrator

Publisher

W. Stephen Comstock

Contents

Preface to the Second Edition

Since its initial publication, *Best Practices for Datacom Facility Energy Efficiency* has been carefully reviewed for necessary updates. Just as Standard 90.1 is under continuous maintenance, the publications of Technical Committee (TC) 9.9 continue to be modified to include the most current industry thinking and consensus.

The primary changes in this second edition center on the updated environmental envelope and relate to the recommended temperatures at the inlets of the equipment operating in datacom facilities. These changes were approved by TC 9.9 during the 2008 ASHRAE Annual Conference in Salt Lake City and are the basis for the second edition of the *Thermal Guidelines for Data Processing Environments*. In addition, the 17 original equipment manufacturers on the committee have agreed to the revised temperature rate-of-change specification, and those changes are also reflected here.

An Appendix G has been added to this second edition of *Best Practices for Datacom Facility Energy Efficiency* that reflects the changes to *Thermal Guidelines*. Tables 2.1 and C.1 and Figures 2.2a, 4.3a, and 4.5a have also been updated. Finally, some typos and errors in reference information have been corrected.

Acknowledgments

The information in this guide was produced with the help and support of the corporations listed below.

American Power Conversion

ANCIS

ANSYS/Fluent,Inc.

Bell South

Chatsworth Products, Inc.

Cisco

Citigroup

Data Aire, Inc.

Degree Controls

Dell Computers

Department of Defense

DLB Associates Consulting Engineers

EYP Mission Critical Facilities

Fluent, Inc.

Fujitsu Laboratories of America

Heapy Engineering

Hewlett Packard

IBM

Intel Corporation

Lawrence Berkeley National Laboratory

Liebert Corporation

Lytron

Oracle

Panduit

Rice University

Rittal

SGI

Spraycool

Stulz-ATS

Sanmina-SCI

Sun Microsystems

Syska & Hennessy Group, Inc.

Unisys

Uptime Institute

Vette Corp.

Wright Line, LLC

York

ASHRAE TC 9.9 wishes to particularly thank the following people:

- **Tom Davidson, Magnus K. Herrlin, Kishor Khankari, Doug McLellan, Izuh Obinelo, Michael K. Patterson, Annabelle Pratt, Joe Prisco, Roger Schmidt, Vali Sorell,** and **Dan Sullivan** for their participation as chapter leads, which included numerous conference calls, writing, and review.

- **Dr. Roger Schmidt** of IBM, Chair of TC9.9, for his invaluable participation in the writing and final editing of this book.

- **Mr. Tom Davidson** of DLB Associates Consulting Engineers for his overall leadership, as well as editing of multiple drafts, of this book.

- **Mr. Don Beaty** of DLB Associates Consulting Engineers, former Chair of TC9.9, for his vision for this book and for his drive and leadership in turning that vision into a reality.

In addition, ASHRAE TC 9.9 would like to thank the following people for their substantial contributions to the creation of this book: John Bean, Christian Belady, Deva Bodas, Kevin Bross, Eric Burger, Tahir Cader, Herman Chu, David Copeland, Mark Germagian, Bill Hay, Rhonda Johnson, Richard Jones, Mukesh Khattar, Geoffrey Lawler, Chris Malone, Timothy McCann, David Moss, Shlomo Novotny, Rick Pavlak, Terry Rodgers, Jeffrey Rutt, Mike Scofield, Grant Smith, Fred Stack, Ben Steinberg, Robert Sullivan, Jeff Trower, William Tschudi, Jim VanGilder, Herb Villa, Alex Vukovic, Kathryn Whitenack, and Lang Yuan.

Second Edition

ASHRAE TC 9.9 would like to thank the following people for their work on this important new environmental envelope (see Appendix G) for improving increased energy savings in data centers: David Moss, Dell; David Copeland, Sun Microsystems; Tim McCann, SGI; Bill French, EMC; Hermann Chu, Cisco Systems; Mike Bishop, Nortel; Chris Malone and Glenn Simon, Hewlett-Packard; Jim Nicholson, AMD; Greg Pautsch, Cray; William Ling, Lucent Technologies; Victor Chiriac, Freescale Semiconductor; Grant Smith, Unisys; Roger Schmidt, IBM; Leo Volpe, Hitachi Global Storage Technologies; and Jonathan Kellen, Seagate Technology, for their active participation, including numerous conference calls, writing/editing, and review.

In addition, ASHRAE TC 9.9 wishes to thank the following people: Vali Sorell, Syska Hennessy Group; Bob Blough, Emerson Network Power; Nick Gangemi, Data Aire; Rhonda Johnson, Panduit; and Alan Claassen and Hussain Shaukatullah, IBM Corp.

1

Introduction and
Best Practices Summary

1.1   PURPOSE

Sustainable design, global warming, depleting fuel reserves, energy use, and operating cost are becoming increasingly more important. These issues are even more important in datacom equipment centers for reasons such as the following:

*   Large, concentrated use of energy (can be 100 times the watts per square foot of an office building).
*   Operations running 24 hours, 7 days a week, have about 3 times the annual operating hours as other commercial properties.

The intent of this publication, as part of the ongoing ASHRAE Datacom Series, is to provide the reader with detailed information on the design of datacom facilities that will aid in minimizing the life-cycle cost, and to maximize energy efficiency in a facility to align with ASHRAE's stated direction (from the 2006 Strategic Plan) to "lead the advancement of sustainable building design and operations."

1.2   BACKGROUND

1.2.1   Historical Perspective and Trends

Figure 1.1 shows a historical picture of a computer center. The first computers were typically quite large, and consisted of a core processing machine with peripherals spaced around a (typically large) room providing storage, input/ output (I/O), and other functions. The power density was not as high as the present, but from a facilities perspective the cooling and conditioning of information technology (IT) equipment was still a challenge. Many of the early mainframes were liquid cooled. Peripheral devices also generated heat, and reliability of tape drives was especially susceptible to the rate of change of both temperature and humidity. To meet the cooling needs of computer rooms, the first computer room air-conditioning (CRAC) unit was developed and shipped in 1955. Since then,

Figure 1.1 Historical data center with mainframe computer and tape storage.

Figure 1.2 Sample motherboard of a modern server.

a variety of methods have evolved within the datacom industry to provide the necessary cooling of the datacom equipment.

The advent of the Internet has created a significant and growing market for computer servers, which perform many of the same functions of the mainframe equipment shown in Figure 1.1, but on a microscale. Figure 1.2 shows a representative motherboard of a server. The efficiency of servers (in terms of computations/ watt) has increased steadily over the years, but miniaturization has occurred at a faster rate than the increase in energy efficiency, resulting in higher power densities at all levels of the datacom equipment package as well as at the facility housing the datacom equipment.

ASHRAE has documented these power density trends in previous publications, and a chart that estimates trends through 2014 based on kW per rack is reproduced here as Figure 1.3. More detailed information on these trends can be obtained from the ASHRAE publication, *Datacom Equipment Power Trends and Cooling Applications* (ASHRAE 2005c).

In order to perform a serious analysis of energy efficiency, the first step is benchmarking existing energy consumption. Lawrence Berkeley National Laboratory (LBNL) initiated a benchmarking study of data centers in 2001, a study that eventually evolved to examine energy use in 22 data centers (LBNL 2007c). An interesting result of the study was the wide variation in the ratio of the amount of power consumed by the IT equipment to the amount of power provided to the entire facility. This ratio varied between 33% and 75%, with the remaining power used by HVAC equipment, uninterruptible power supply (UPS), lighting, and other loads. Average power allocation was difficult to assess because of variation in site characteristics, but an approximate average allocation of 12 of the 22 data centers benchmarked is shown in Figure 1.4. The types of equipment found in each category are listed below:

- HVAC cooling: chillers, pumps, dry coolers, condensers, and/or cooling towers.
- HVAC fans: fan energy in HVAC equipment (for sites with compressorized CRACs; some cooling energy may also be included in this category).
- Lighting: space lighting.
- UPS: uninterruptible power supply equipment losses.
- Servers: a catch-all term for the IT equipment (servers, storage, I/O, and networking equipment) on the datacom equipment floor performing the intended function of the datacom facility.
- Other: a catch-all term for all items that do not fall into the above categories; this portion of the energy pie might include miscellaneous electrical losses, a small support office area, etc., depending on the exact facility.

The two most important justifications for this publication are listed below.

1. *End user's "bottom line."* The cost of building and operating datacom facilities is significant, and in a commercial market must be justified by a return on investment. One of the ramifications of miniaturization of datacom equipment, as previously noted, has been the increase in power density of datacom equipment.

A parallel trend, however, has been that the cost of datacom equipment per watt of input power has been decreasing, allowing end users to purchase more "watts" of equipment for a given investment cost. This can have a significant impact on total cost of ownership (TCO). *TCO* is a term used in

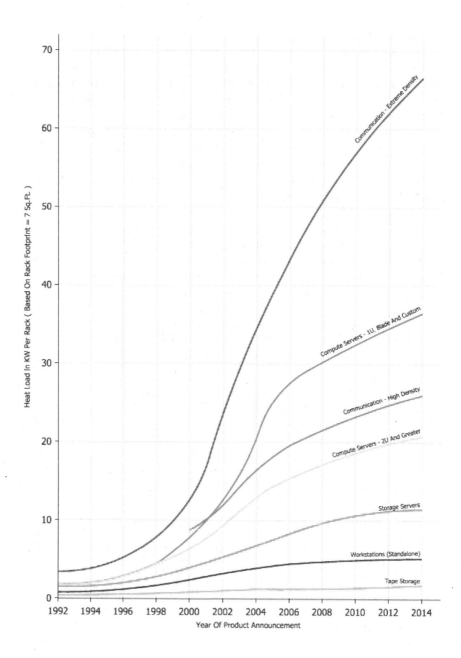

Figure 1.3 ASHRAE power trend chart (ASHRAE 2005c).

Average Data Center Power Allocation

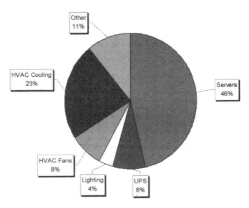

Figure I.4 Average power allocation for 12 benchmarked data centers (LBNL 2007a).

the datacom industry; the building industry uses *life-cycle costing*. Figure 1.5 provides a comparison of server costs and energy costs for powering the servers, in terms of annual expenditures for the industry. Other industry sources indicate that the combination of infrastructure and energy costs exceeds the server costs for a datacom facility (Belady 2007; Sullivan 2007). Since the infrastructure/energy cost is an increasing component of TCO, a strong emphasis needs to be placed on this topic to keep a datacom facility energy efficient and at the lowest cost to support the level of reliability and availability of the equipment it houses.

2. *National Energy Policy.* The interest of the US Federal Government in promoting energy efficiency has been apparent since the energy crisis of the early 1970s. Interest in computer data efficiency started with EPA ENERGY STAR® ratings of personal computer equipment in June 1992. The EPA published an ENERGY STAR Specification for Enterprise Computer Servers in December 2006. Standards for desktop and workstation efficiency became effective in July 2007. The purpose of US Public Law #109-431, passed and signed in December 2006, is to "study and promote the use of energy efficient computer servers in the United States." The scope of PL109-431, however, extended beyond the server equipment to essentially include all aspects of data center efficiency.

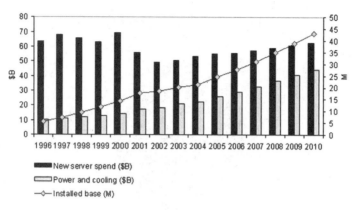

Source: IDC, Worldwide Server Power and Cooling Expense 2006–2010 Forecast, Doc #203598, September 2006.

Figure I.5 Comparative forecast of annual server power and cooling expenditures to new server expenditures, through 2010.

The overall justification for this document is thus one of minimizing TCO for the end user, helping to conserve energy resources on a national (and global) scale, and alignment with ASHRAE's sustainability goals, as indicated in Section 1.1.

1.2.2 Energy Efficiency and TCO Definitions

This book provides basic definitions and explanations of most terms in the glossary. Discussion of two basic terms, however, is included in this chapter, as the terms are used throughout the book. These terms are *efficiency* and *total cost of ownership* (TCO).

Efficiency

The term *efficiency* or *energy efficiency* can be defined in hundreds of ways to correspond to various applications, so it is important to establish definitions that are of value in datacom facility analysis.

The basic ASHRAE definition of *efficiency* is "the ratio of the energy output to the energy input of a process or a machine" (ASHRAE 1991). If this definition is applied to a data center (or some component thereof) the process has to be defined along with definitions of the energy output and energy input of the process. Various entities have tried to follow this approach for data centers, and a few definitions have evolved.

Cooling Plant Efficiency. A major component of the energy efficiency of a data center is the efficiency of the cooling plant. If facility cooling is the process, one could define the IT load (kW) of the facility as the energy output, and define the energy use (kW) of the cooling plant as the energy input. This definition of cooling plant efficiency (computer load kW output / cooling plant kW input) is currently used by Lawrence Berkeley National Laboratory (in an inverted form) as a metric for a series of benchmarking studies (LBNL 2007c), and can serve as a good marker of cooling plant efficiency. This ratio has wide variance based on the referenced benchmarking reports, and detailed analysis of the sources of input energy can yield excellent information for improving energy efficiency. Chapters 2 through 5 of this book are dedicated to a discussion of parameters relevant to cooling plant efficiency, though other chapters are also integral to a complete analysis.

Electrical Distribution Efficiency. A look solely at cooling plant efficiency can allow one to miss another source of datacom facility inefficiency—the electrical distribution process. Datacom facilities may have transformers, generators, uninterruptible power supplies (UPSs), power distribution units (PDUs), and switchgear. Chapter 7 of this book addresses the specific issue of electrical distribution efficiency in more detail.

Datacom Equipment Efficiency. One could justifiably argue that having an efficient cooling plant and an efficient electrical distribution system do not by themselves constitute an energy-efficient data center if the datacom equipment is not also efficient. The definition of *datacom equipment efficiency*, as used in this book, is the ratio of the load energy to the input energy, where the load includes microprocessors and other integrated circuits, volatile memory, and hard disk drives, and the input energy is the energy delivered to the input of the power supply unit (see Section 8.5 for more detail).

Other Efficiencies. There are other sources of power consumption in a data center, such as lighting. Lighting efficiency is covered in *ANSI/ASHRAE/IESNA Standard 90.1-2004, Energy Standard for Buildings Except Low-Rise Residential Buildings* (ASHRAE 2004a), and other publications and is not covered further in this book.

Datacom Facility Efficiency. If the useful output of a datacom facility is defined as power having reached the datacom equipment, and the input is the total electrical power provided to a facility by the electric utility, then an overall efficiency metric for the data center can be established.

The Green Grid, a consortium dedicated to developing and promoting energy efficiency for data centers and information services, has adopted the terms *power usage effectiveness* (PUE) and *datacenter efficiency* (DCE).

The term *power usage effectiveness* (PUE) takes the following form:

$$PUE = \frac{\text{Total Facility Power}}{\text{IT Equipment Power}} \quad \text{(Green Grid 2007)}$$

The reciprocal of the PUE is the term *data center efficiency* (DCE), which takes the following form:

$$DCE = \frac{IT\ Equipment\ Power}{Total\ Facility\ Power}\ (Green\ Grid\ 2007)$$

As an example, the DCE of an average data center, as depicted in Figure 1.4, would be the ratio of the server power to the total facility power, or 46%. In using an overall definition, one has to determine if the ratio is a snapshot of efficiency or a more comprehensive value that could reflect, as an example, average annual data center efficiency. This topic will be addressed later in the book. Definitions similar to DCE are in use by other organizations.

One needs to be careful, when using any metric, that an "apples-to-apples" comparison is being made. For instance, the above definition includes fan power within the server as "IT equipment power," and fan power within air-handling units and CRAC units as part of "total facility power." Innovations that shift the fan power from the servers to the air-handling units, however, will lower DCE, but may also result in a concurrent decrease in total facility power.

Combined Datacom Facility and Datacom Equipment Efficiency. As indicated in Chapter 8, the term *energy efficiency* as discussed in this book should not be confused with the measure of *performance per watt*, which is the ratio of the performance of the equipment, measured in operations per second or in the time it takes to complete a specific benchmark, to the average input power of the equipment. An end user may want to make a comparison of this as part of an overall TCO analysis of a data center, but a full analysis (i.e., benchmark datacom equipment performance per watt of input power to the portion of the facility dedicated to the datacom equipment and its mechanical and electrical support infrastructure) extends beyond the full technical scope of this book.

Total Cost of Ownership (TCO)

TCO is a financial estimate (originating in the datacom industry) designed to assess direct and indirect costs related to the purchase of any capital investment. A TCO assessment ideally offers a final statement reflecting not only the cost of purchase but all aspects of further use and maintenance of the equipment, device, or system considered. This includes the costs of training support personnel and the users of the system, costs associated with failure or outage (planned and unplanned), diminished performance incidents (i.e., if users are kept waiting), costs of security breaches, costs of disaster preparedness and recovery, floor space, energy, water, development expenses, testing infrastructure and expenses, quality assurance, incremental growth, decommissioning, and more. TCO analysis can be used to evaluate alternatives ranging from complete datacom facilities to a piece of HVAC or IT equipment in that data center, and provides a cost basis for determining the economic value of that investment.

1.2.3 Overview of ASHRAE

The American Society of Heating, Refrigerating and Air-Conditioning Engineers (ASHRAE), which was founded in 1894, is an international nonprofit technical engineering society comprised of around 55,000 members. ASHRAE fulfills its mission of advancing HVAC&R to serve humanity and promote a sustainable world through research, standards writing, publishing, and continuing education.

ASHRAE has around 100 standard and guideline project committees that establish recommended design and operation practice, and is one of only five standards-developing organizations in the United States that can self-certify that its standards have followed the American National Standards Institute's (ANSI's) standards development procedures.

ASHRAE has some 100 technical committees that drive the ASHRAE research program; develop standards; sponsor the technical program at ASHRAE meetings; develop technical articles, special publications, and educational courses; and write the *ASHRAE Handbook*.

ASHRAE's government affairs program provides a critical link between ASHRAE members and government through contributing technical expertise and policy guidance to Congress and the Executive branch. Current priorities include energy efficiency; building codes; science, technology, engineering, and mathematics education; indoor environmental quality; and building security.

1.2.4 Overview of ASHRAE Technical Committee (TC) 9.9

Key technical experts of the major IT manufacturers recognized that power and cooling capacities were going to become increasingly more challenging for the industry. Further, they saw no vendor neutral professional society holistically addressing the technical aspects of the data center industry. They were also seeing an increasing need for the collaboration and coordination of the IT industry and the facilities industry.

Due to ASHRAE's major international presence and leadership, long history, and major publishing infrastructure (including model codes, standards, guidelines, courses, etc.), the IT manufacturers saw ASHRAE as the source to publish unbiased information. As a result, Roger Schmidt (IBM) and Don Beaty (DLB Associates) started the formal process of providing ASHRAE the justification for creating a dedicated technical committee for data center facilities.

Since no other vendor-neutral, nonprofit organization existed for data center facilities, the committee was organized and its members carefully selected to address the broadest possible scope. For example, even the committee title, "Mission Critical

Facilities, Technology Spaces, and Electronic Equipment," reflects a broad perspective (facility, or macro, down to the electronics, or micro).

TC 9.9 members include experts from the IT manufacturers as well as the facility design, construction, and operation areas. The committee also includes members from numerous countries around the world to help provide even a broader perspective. A number of these committee members are not members of ASHRAE and are not thermal engineers.

The focus of the committee is to identify informational and technical needs of the data center industry and to meet those needs. Where the committee does not have the full range of resources or expertise, resources are sought and added to the team. These needs in some cases are not HVAC based, so the committee and ASHRAE's publishing capabilities are employed as a means of meeting the industry's needs.

To summarize, TC 9.9 has the following major objectives:

- Produce unbiased technical material on data center HVAC
- Provide unbiased training on data center HVAC
- Provide a forum for publishing unbiased technical material on subjects other than HVAC for the data center industry

1.2.5   Overview of ASHRAE TC 9.9 Datacom Book Series

The ASHRAE Datacom (data centers and communication facilities) Series is ASHRAE TC 9.9's primary means to meet the informational needs of the data center industry. The content is intended to provide value to *both* technical and nontechnical readers.

The books vary in that sometimes they are totally independent of previous books in the series, while occasionally they may build on previous books in the series.

At the time of publication of this book, the following books have been published (additional books are well underway):

- *Thermal Guidelines for Data Processing Environments* (ASHRAE 2009)
- *Datacom Equipment Power Trends and Cooling Applications* (ASHRAE 2005c)
- *Design Considerations for Datacom Equipment Centers* (ASHRAE 2005d)
- *Liquid Cooling Guidelines for Datacom Equipment Centers* (ASHRAE 2006c)
- Structural and Vibration Guidelines for Datacom Equipment Centers (ASHRAE 2007j)

1.2.6   Primary Users for This Book

Those involved in the design, construction, commissioning, operating, implementation, and maintenance of datacom equipment centers can all benefit from this book. In addition, those who develop and design electronic, cooling, and other infrastructure equipment will benefit from these guidelines. Specific examples of users of this document would include the following:

- Computer equipment manufacturers—research and development engineers and marketing and sales organizations
- Infrastructure equipment manufacturers—cooling and power
- Consultants
- General construction and trade contractors
- Equipment operators, IT departments, facilities engineers, and chief information officers

Energy efficiency is an important metric because energy usage in telecommunications facilities and data centers (the term *datacom facilities* is used in this book to cover both) is a significant portion of facility operating costs. Likewise, TCO encompasses an extremely important set of parameters that, analyzed as a whole, can minimize costs and increase the overall efficiency of datacom facilities.

1.3   BEST PRACTICES AND CHAPTER SUMMARIES

1.3.1   Best Practices

Table 1.1 is intended to provide summaries of the information discussed in detail in the remainder of this book.

Table 1.1   Best Practices

Environmental Criteria (Chapter 2)

1. Colder does not mean better; adoption of ASHRAE temperature and humidity ranges in Table 2.1 can achieve greater equipment efficiency and an increase in economizer hours.
2. Implement control strategies that eliminate cooling units fighting (one heating while another is cooling).
3. Size filters for a low pressure drop (less energy).
4. Maintain filters to avoid increased pressure drop (more energy).

Mechanical Equipment and Systems (Chapter 3)

1. Choose components and systems that operate efficiently during part- and full-load conditions.
2. Develop several alternatives and the associated TCO for each alternative.
3. Model cooling plant operation for entire year for both current operations and hypothetical future upgrades.
4. Use load matching techniques, such as variable-speed drives and variable capacity compressors.
5. Chillers often account for a significant amount of cooling energy in chilled-water systems; strategies to minimize chiller energy consumption should be investigated in detail.

Table 1.1   Best Practices *(continued)*

---

6. For CRAC units, focus the cooling solution on very high sensible/total cooling capacities per the revised ANSI/ASHRAE Standard 127-2007 (ASHRAE 2007b).

Economizer Cycles (Chapter 4)

1. Economizer cycles provide an opportunity for substantial energy and cost savings in data centers, but savings and TCO are quite dependent on climate.

2. Water-side economizers generally have lower annualized utilization (except in dry climates), but dehumidification and humidification are not constraints to water-side economizer use and/or efficiency.

3. Adiabatic air-side economizers have the highest potential annualized utilization, but market share is low.

4. Integrated economizer controls should be utilized to allow partial use of the economizer cycle and increase annualized utilization (true for all types of economizers).

5. Raising the supply air setpoint in a facility can significantly increase the number of cooling hours in economizer mode.

Airflow Distribution (Chapter 5)

1. Consider several airflow architectures including overhead, in floor, or hybrid systems.

2. Use computational fluid dynamics (CFD) modeling and associated validation to understand layout configurations.

3. Provide good separation between supply and return air pathways (e.g., hot aisle/cold aisle). When possible locate the down-flow CRAC units at the end of the hot aisle.

4. Select energy-efficient fans, fan motors, and variable-frequency drives to maximize fan efficiency. Design system pressure drop should be minimized.

5. Select as high a supply air temperature as will provide sufficient cooling.

6. Recognize that datacom equipment loads will change over the next 10 to 15 years. Develop a cooling distribution strategy that can adjust to these changes.

HVAC Controls and Energy Management (Chapter 6)

1. Investigate the costs/benefits of different methods for humidity control. System design and control algorithms should allow the primary cooling coils to "run dry," and thus allow for chilled-water reset at light loads without impacting relative humidity.

2. Develop an efficient part-load operation sequence.

3. Use effective variable-speed control and variable capacity control.

4. Consider thermal storage for uninterruptible cooling and stabilizing cooling system (energy savings are also possible but detailed analysis is required).

Electrical Distribution System (Chapter 7)

1. Choose UPSs that operate with high efficiency throughout their expected operating range.

Table 1.1 Best Practices (continued)

---

2. Redundancy should be used only up to the required level; there may be an efficiency penalty for additional redundancy.

3. Select high-efficiency transformers, consider harmonic minimization transformers, but do not compromise other electrical issues.

4. Limit conductor runs by installing PDUs as close to the rack load as possible.

5. Consider distributing high voltage AC or DC power to the point of use.

Datacom Equipment Efficiency (Chapter 8)

1. Replace older equipment with more efficient designs.

2. Install high- efficiency power supplies with power factor correction.

3. Select power equipment from the highest input voltage available within its input voltage rating range.

4. Effectively utilize power management features.

5. Provide variable-speed fans, optimize their control, and effectively interface with the cooling service.

6. Avoid using power supplies rated for much higher power than expected on the platform ("right size").

7. Use only the level of redundancy required to meet the availability requirements.

8. Employ virtualization/consolidation.

Liquid Cooling (Chapter 9)

1. Compare and effectively choose the best cooling fluid for the application.

2. Minimize pumping power.

3. Optimize heat exchanger selection.

4. Consider the use of a cooling distribution unit (CDU) to isolate the liquid cooling loop from the building chilled-water cooling loop. This allows the liquid cooling loop temperature to be set above the room dew-point temperature, thus eliminating condensation.

5. Recognize the datacom equipment loads will change over the next 10 to 15 years. Develop a liquid cooling strategy that can adjust to these changes.

Total Cost of Ownership (TCO) (Chapter 10)

1. Generate a list of valid TCO options, based on research, discussion with experts, etc.

2. Determine the discount rate and time frame for the analysis.

3. Use energy modeling software to provide sufficient data to develop an effective TCO.

Commissioning

1. Establish Owner's Program Requirements (OPRs) in accordance with *ASHRAE Guideline 0, The Commissioning Process* (ASHRAE 2005e).

2. Verify and document that the facility, its systems, and assemblies are designed, installed, and maintained in accordance with the OPRs.

1.3.2   Overview of Chapters

Chapter 1, Introduction. The introduction states the purpose/objective of the publication, some definitions, and includes a brief overview of the chapters.

Chapter 2, Environmental Criteria. The environmental conditions both inside and outside a facility have a significant impact on the choice of an energy-efficient cooling system. The goals of this chapter are to provide an overview of design conditions inside a data center (as published in *Thermal Guidelines for Data Processing Environment*s [ASHRAE 2009]), to compare these conditions to previous criteria listed in ASHRAE, and to indicate ways in which the new criteria can result in increased energy efficiency. The potential impact of climatic conditions on data center energy consumption is also discussed.

Chapter 3, Mechanical Equipment and Systems. The intent of this chapter is to look at the equipment and systems that provide cooling and humidity control for a facility, and to summarize pertinent design considerations to aid in the selection of energy-efficient systems. One could consider this a mini "Systems and Equipment Handbook" for energy-efficient data center environmental conditioning. Types of equipment covered include chillers, CRAC units, humidification equipment, and heat rejection equipment. Types of systems include air-cooled, water-cooled, and evaporative-cooled systems.

Chapter 4, Economizers. The intent of this chapter is to discuss economizer cycles in the various ways that they can be used to save energy in datacom facilities. The chapter includes discussions on both outdoor air economizer cycles and water economizer cycles.

Chapter 5, Airflow Distribution. The air distribution design in a data center is nontrivial, and a substantial percentage of energy costs in an air-cooled data center are associated with air distribution. The intent of this chapter is to identify air distribution parameters that affect energy consumption, and to discuss ways in which energy efficiency can be increased. Among the topics covered are methods of airflow delivery, effect of supply air temperatures, airflow within racks, leakage, and airflow optimization.

Chapter 6, Controls and Energy Management. Control and energy management systems are a key component to energy-efficient operation of a data center. This chapter covers a number of aspects of control systems that have an impact on energy-efficient operations. These include control system architecture, energy-efficient IT equipment control, part-load operation, fans, pumps and variable-speed drives, humidity control, and outdoor air control.

Chapter 7, Electrical Distribution Equipment. Recent studies have identified that the electrical distribution systems in data centers can be a significant source of energy loss. As such, the goal of this chapter is to identify these losses, and to discuss approaches that can increase the efficiency of electrical distribution. The chapter starts with a brief discussion of utility generation, then continues with facility distribution, including an AC and DC overview. Also included are distribution paths; energy-efficiency options, such as reduction in amperage and harmonics; and component efficiencies of UPSs and transformers.

Chapter 8, Datacom Equipment Efficiency. Datacom equipment is continuing to become more energy efficient, measured on the basis of several parameters. There are more efficient power supplies and voltage regulators, lower power processors, and increasingly sophisticated power management. Since datacom equipment needs electrical distribution and cooling infrastructure to support the resulting heat load, efficiency improvements at this base level will also reduce energy use in the distribution system and mechanical plant. This chapter discusses these and many other issues related to datacom equipment energy efficiency.

Chapter 9, Liquid Cooling. Liquid cooling equipment and installations are becoming more common, primarily in response to the increasing power density of datacom equipment and the need to provide cooling infrastructure to this equipment in a compact way. This chapter provides a look at liquid cooling from the standpoints of energy and TCO.

Chapter 10, Total Cost of Ownership. TCO is a broader tool that can be used to design datacom facilities in an efficient and cost-effective manner. This chapter starts by defining TCO, and discusses its applicability for making design decisions in the datacom industry. The capital and operational components of a TCO model are discussed, and advice on what to include in comparative TCO models is provided.

Chapter 11, Emerging Technologies and Future Research. In this chapter, a listing of some emerging technologies that could have an impact on energy efficiency in datacom facilities is presented, and suggestions for future research are made.

The book concludes with a bibliography/reference list and the following appendices: Appendix A, Glossary of Terms; Appendix B, Facility Commissioning, Operations, and Maintenance; Appendix C, Telecom Facility Experience; Appendix D, Applicable Codes and Standards; Appendix E, Sample Control Sequences; and Appendix F, SI Units for Tables and Figures.

2

Environmental Criteria

2.1   INTRODUCTION

Just as the design temperature in an office building will impact HVAC equipment sizing and annual energy consumption, the design criteria for a data center will have a similar impact on cooling plant energy consumption. In looking at the overall usage in a facility, the environmental criteria will impact energy consumption in the portions of the Figure 2.1 pie chart labeled as "HVAC Cooling" and "HVAC Fans." "HVAC Cooling" generally encompasses all of the equipment used to cool and condition a datacom facility. "HVAC Fans" generally encompasses the fan energy used to move air within a datacom facility, as well as outdoor air, but does not include the fans that are integral to the datacom equipment. Taken together, these two slices of the energy

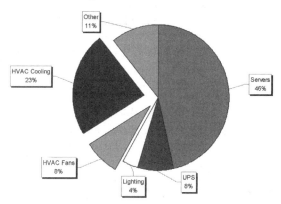

Average Data Center Power Allocation

Figure 2.1   Energy consumption impact of environmental criteria (LBNL average of 12 data centers.

17

pie account for 31% of the energy consumption of an average data center (LBNL 2007a). The level of impact of the environmental conditions on energy consumption of the HVAC cooling and HVAC fans systems will depend on several parameters, including ambient conditions and system economizer configuration.

The flow of this chapter will be to introduce the generally accepted environmental classes for datacom facilities. A discussion of the impact of the environmental conditions corresponding to these classes on datacom energy use follows. The chapter concludes with a summary and list of major recommendations.

2.2   OVERVIEW OF ENVIRONMENTAL CLASSES

Environmental requirements of datacom equipment may vary depending on the type of equipment and/or manufacturer. However, there has been agreement among a consortium of manufacturers on a set of four standardized conditions (Classes 1–4). These conditions are listed in *Thermal Guidelines for Data Processing Environments* (ASHRAE 2009). An additional classification, the Network Equipment—Building Systems (NEBS) class, is typically used in telecommunications environments. This publication is primarily concerned with TCO and energy use in Class 1, Class 2, and telecom. Definitions of these classes are provided below.

- **Class 1:** Typically a datacom facility with tightly controlled environmental parameters (dew point, temperature, and relative humidity) and mission critical operations; types of products typically designed for this environment are enterprise servers and storage products.
- **Class 2:** Typically a datacom space or office or lab environment with some control of environmental parameters (dew point, temperature, and relative humidity); types of products typically designed for this environment are small servers, storage products, personal computers, and workstations.
- **NEBS:** Per Telcordia GR-63-CORE (Telecordia 2002) and GR-3028-CORE (Telecordia 2001), typically a telecommunications central office with some control of environmental parameters (dew point, temperature and relative humidity); types of products typically designed for this environment are switches, transport equipment, and routers.

Class 3 and 4 environments are not designed primarily for the datacom equipment and are not addressed in this book. For further information on these classes, reference should be made to *Thermal Guidelines for Data Processing Environments* (ASHRAE 2009).

2.3   ENVIRONMENTAL CRITERIA

Table 2.1 lists the recommended and allowable conditions for the Class 1, Class 2, and NEBS environments as defined by the sources that are footnoted. Figure 2.2a shows recommended temperature and humidity conditions for Class 1,

Table 2.1   Class 1, Class 2, and NEBS Design Conditions

Condition	Class 1/Class 2		NEBS	
	Allowable Level	Recommended Level	Allowable Level	Recommended Level
Temperature control range	59°F–90°F[a,f] (Class 1) 50°F–95°F[a,f] (Class 2)	64.4°F–80.6°F[a,g]	41°F–104°F[c,f]	65°F–80°F[d]
Maximum temperature rate of change	9°F. per hour[a]		2.9°F/min.[d]	
RH control range	20%–80% 63°F. max dew point[a] (Class 1) 70°F. max dew point[a] (Class 2)	41.9°F dew point– 60% RH and 59°F dew point[a, g]	5%–85% 82°F max dew point[c]	Max 55%[e]
Filtration quality	65%, min 30%[b] (MERV 11, min MERV 8)[b]			

[a]These conditions are inlet conditions recommended in *Thermal Guidelines for Data Processing Environments, Second Edition* (ASHRAE 2009).
[b]Percentage values per the ASHRAE Standard 52.1 dust-spot efficiency test. MERV values per ASHRAE Standard 52.2. Refer to Table 8.4 of *Design Considerations for Datacom Equipment Centers* (ASHRAE 2005d) for the correspondence between MERV, ASHRAE 52.1, and ASHRAE 52.2 filtration standards.
[c]Telecordia 2002 GR-63-CORE.
[d]Telecordia 2001 GR-3028-CORE.
[e]Generally accepted telecom practice. Telecom central offices are not generally humidified, but grounding of personnel is common practice to reduce ESD.
[f]Refer to *Thermal Guidelines for Data Processing Environments, Second Edition* (ASHRAE 2009) for temperature derating with altitude.
[g]68°F/40% RH corresponds to a wet-bulb temperature of 54°F. *Caution:* operation of DX systems with a *return* wet-bulb temperature below 54°F has a likelihood of causing freezing coils.

Class 2, and NEBS classes on a psychrometric chart. For comparison purposes it also shows the recommended conditions as listed in the *2003 ASHRAE Handbook— Applications* (ASHRAE 2003). Figure 2.2b shows allowable temperature and humidity conditions for the same classes. It should be noted that the dew-point temperature is also specified, as well as the relative humidity. The stated conditions correspond to conditions *at **the inlet** to the datacom equipment* (typically the cold aisle), not to "space conditions" or to conditions at the return to the air-conditioning equipment. Temperatures in other parts of the datacom spaces, such as hot aisles, can exceed the recommended (and even allowable) conditions set forth in Table 2.1. Environmental conditions for ancillary spaces (e.g., battery storage and switchgear

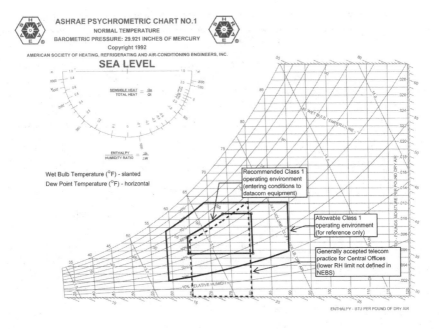

Figure 2.2a Recommended data center Class 1, Class 2, and NEBS operating conditions. Refer to Appendix F for an equivalent figure with SI units.

rooms) do not necessarily require the stringent environmental conditions of Class 1 and Class 2.

2.3.1 Temperature

Equipment exposed to high temperature, or to high thermal rate of change, can experience thermal failure particularly when repeatedly exposed to high thermal gradients. Required inlet air conditions to datacom equipment should be checked for all equipment, but a typical recommended range is 64.4°F to 80.6°F (18°C to 27°C) (ASHRAE 2009). For telecommunications central offices and NEBS, the ranges are almost identical (see Table 2.1). The recommended range of 65°F to 80°F (18°C to 27°C) in GR-3028-CORE is based on a TCO analysis (Herrlin 1996), including the following:

- Energy costs associated with operating the HVAC system
- Battery replacement costs
- Datacom equipment repair
- Cleaning costs

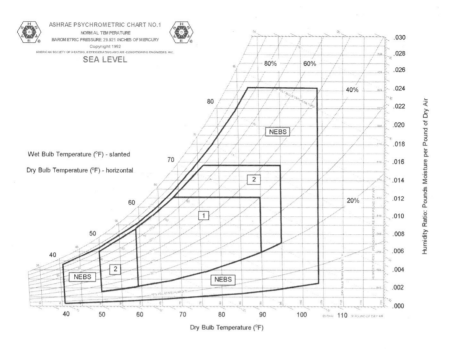

Figure 2.2b Allowable data center Class 1, Class 2, and NEBS operating conditions. Refer to Appendix F for an equivalent figure with SI units.

Human comfort is typically considered a secondary consideration in data centers, since the number of people working in the computer-dominated sections of data centers is small. Allowing data centers to operate with less critical attention to human comfort can allow them to operate more efficiently. For instance, the hot-aisle/cold-aisle airflow arrangement is generally considered to be energy efficient, but can result in conditions that are somewhat warmer than the comfort zone in the hot aisle. For reference, it is noted that OSHA guidelines allow for continuous light work duty with space temperatures as high as 115°F (46°C) if the space absolute moisture is maintained no greater than 58.6 grains/pound (8.25g/kg) of dry air, equivalent to 72°F (22°C) at 50% RH (OSHA 2007). Data center operators and designers are encouraged to consult all local and relevant codes regarding worker heat stress, and adjust values accordingly.

There are three primary energy implications related to the temperature of a datacom facility. All three of these affect the "HVAC Cooling" piece of the pie chart shown in Figure 2.1. These implications are as follows:

- Vapor compression cycle efficiency
- Impact of temperature on HVAC system economizer hours
- Impact of operating dead band

Vapor-Compression Cycle Efficiency. Thermodynamic efficiency of vapor-compression cycles can be increased: raising the space temperature allows the vapor-compression cycle to run with a smaller temperature differential. This increases the thermodynamic efficiency of the cycle. With chilled-water plants, the rule-of-thumb increase in efficiency is 1%–3% for every 1°F (0.6°C) rise in evaporator temperature. Chiller part-load efficiencies in the range of 0.20 kW/ton (COP = 17.6) are possible with 60°F (16°C) chilled water and concurrent low condenser water temperature. For comparison, the current ASHRAE standard for a large centrifugal chiller at standard rating conditions is 0.576 kW/ton (COP = 6.1) (ASHRAE 2004a). In many climates, it may be possible to produce 60°F (16°C) water through evaporative cooling without the use of chillers for many hours of the year.

HVAC System Economizer Hours. As the supply temperature to the inlet of datacom equipment is raised, it typically means that the HVAC equipment can operate in "economizer" or "free cooling" mode for a greater number of hours per year. For instance, maintaining a datacom facility at 75°F (24°C) will almost certainly result in a greater number of economizer hours than a facility with a 68°F (20°C) temperature. Chapter 4, "Economizers," is devoted to this topic, so it will not be discussed further here, but the savings can be considerable.

Some European telecommunications firms have been able to eliminate refrigerant-based cooling systems either most of the time or entirely by (aggressively) allowing the space temperature and humidity to vary through the full range of allowable equipment operating environments. These facilities utilize 100% outdoor air (direct free cooling) some or all of the time, in combination with detailed space environmental design using CFD techniques. The environmental system energy usage of these facilities is claimed to be less than 50% of a system utilizing refrigerant-based cooling (Cinato et al. 1998; Kiff 1995).

HVAC System Operating Dead Band. A dead band in a control system is the area of a signal range where no action occurs (i.e., the system is dead). In reference to datacom facilities, this refers to the acceptable temperature range over which neither heating nor cooling is required. The recommended operating temperature dead band for Class 1 and Class 2 facilities is between 64.4°F to 80.6°F (18°C to 27°C), corresponding to a range of 16.2°F (9°C). This is more than twice the dead band of 4°F (2.2°C) recommended for a legacy datacom facility (ASHRAE 2003). This larger dead band can allow for less "fighting" among adjacent HVAC units that may have both heating and cooling capability.

ASHRAE Standard 90.1 (ASHRAE 2004a) was recently amended to rescind an exemption provided to data centers in regard to dead bands (ASHRAE 2006a). Zone thermostatic controls must now be capable of providing a dead band of at least 5°F

(3°C) within which the supply of heating and cooling energy to the zone is shut off or reduced to a minimum.

2.3.2   Humidity

High relative humidity may cause various problems to datacom equipment. Such problems include conductive anodic failures (CAF), hygroscopic dust failures (HDF), tape media errors and excessive wear, and corrosion. In extreme cases, condensation can occur on cold surfaces of direct-cooled equipment. Low relative humidity increases the magnitude and propensity for electrostatic discharge (ESD), which can damage equipment or adversely affect operation. Tape products and media may have excessive errors when exposed to low relative humidity.

There are three primary energy implications related to the humidity of a datacom facility. All three of these affect the "HVAC Cooling" piece of the pie chart shown in Figure 2.1.These implications are as follows:

- Impact of humidity level on the energy cost of humidification and dehumidification allowed in a facility
- Impact of relative humidity setpoint on dehumidification and reheat costs
- Impact of the humidity dead band on energy cost

Humidification and Dehumidification Setpoint Energy Impact. The current recommended relative humidity range for Class 1 and Class 2 data centers (41.9°F dew point–60% RH and 59°F dew point) should minimize maintenance and operational problems related to humidity (either too low or too high), but there is an energy cost associated with maintaining the environment in this range. Some equipment manufacturers may have wider operating ranges; lowering the low end of the range will result in lower humidification energy costs, and raising the high end of the range will result in lower dehumidification costs. As shown in Figure 2.2a, and also discussed in the Telecom Facility Experiences appendix, the telecom industry has generally not found it economical to humidify their facilities in any climate, though they typically have strong personnel-grounding protocols.

> Savings Example #1: As an example of the cost savings possible from resetting the minimum humidity from 45% to a dew point of 41.9°F, consider the following:

> Assume that a data center requires 10,000 cfm of outdoor air, and that for 50% of the year the air needs to be humidified to either (a) 45% RH or (b) 41.9°F dew point at a temperature of 72°F. Electric source heat is used to provide the humidification, and the average energy cost is $0.10/ kWh.

> Enthalpy of 72.0°F, 45% RH air: 25.51 Btu/lb
> Enthalpy of 72.0°F, 41.9°F dew point: 23.51 Btu/lb

Difference: 2.0 Btu/lb

Mass flow rate: 10,000 cfm × 0.074 lb/ft^3 × 60 min/h = 44,400 lb/h

Energy savings of 41.9°F dew point: 44,400 lb/h × 2 Btu/lb = 88,800 Btu/h

88,800 Btu/h × 1 kW/3412 Btu/h = 26.02 kW

62.02 kW × 4380 h/yr × $0.10/kWh = $11,399/yr

Impact of Relative Humidity Setpoint on Dehumidification and Reheat Costs. Dehumidification in datacom environments often occurs through the use of the cooling coil in CRAC units. Typically the cooling coil temperature is dropped if dehumidification is required, since this increases the amount of condensation on the coil and decreases space absolute humidity. Since the cooler coil can result in overcooling of the space, the colder discharge air is often reheated in constant-volume systems, resulting in high energy use through simultaneous heating and cooling. Increasing the maximum relative humidity level will reduce the number of hours that the cooling coil is operating in the dehumidification mode and, thus, the use of simultaneous heating and cooling. (New ASHRAE energy codes preclude this method of providing humidity control, since it involves simultaneous cooling and heating, but many units in the field still utilize this humidity control technique.)

Operating Dead Band Energy Impact. The legacy relative humidity dead band for data centers has been tight, and is one of the reasons why reheat coils have historically been available in CRAC units to maintain relative humidity in the desired range. The addition of the reheat coil has, in turn, resulted in occasional "fighting" between adjacent CRAC units, with some units cooling and "dehumidifying" while adjacent units heat and/or humidify the air. Maintaining as wide a dead band as possible, or maintaining humidity levels through dew-point control at the facility level, will typically minimize the waste often associated with a tight humidity dead band.

ASHRAE Standard 90.1 was recently amended to rescind an exemption provided to data centers in regard to simultaneous operation of both humidification and dehumidification equipment (ASHRAE 2006a, Addendum h to Standard 90.1). To quote from the amended standard, "Where a *zone* is served by a system or systems with both humidification and dehumidification capability, means (such as limit switches, mechanical stops, or, for DDC [direct digital control] systems, software programming) shall be provided capable of preventing simultaneous operation of humidification and dehumidification equipment."

2.3.3 Filtration and Contamination

Table 2.1 contains, for particulates but not for gaseous contaminants, both recommended and minimum allowable filtration guidelines for recirculated air in a data center. Dust can adversely affect the operation of datacom equipment, so high-quality filtration and proper filter maintenance are essential. Corrosive gases can quickly destroy the thin metal films and conductors used in printed circuit boards, and can also cause high resistance at terminal connection points.

In addition, the accumulation of dust and other contaminants on surfaces needed for heat removal (i.e., heat sink fins) can retard the ability of the heat removal device to perform properly.

The major energy impacts of filtration and contamination to a datacom facility, as listed and described below, fall into the "HVAC Fans" slice of the energy pie. The major impacts are the following:

- Filtration levels required for economizer operation
- Energy loss due to high filter pressure drop and/or poor filter maintenance

Filtration Levels Required for Economizer Operation. Outside air should be treated and preconditioned to remove contamination, salts, and corrosive gases before it is introduced into the data and/ or communications equipment room. This is particularly true for systems that utilize an air-side economizer, since 100% outdoor air will be introduced into the data center at certain times of the year. Higher filtration levels may result in less product failure, but there may be an increased air pressure drop and associated energy use. A decision must be made on the proper level of filtration to minimize facility TCO where an air-side economizer is employed. Research on particulate levels outside and inside a data center has been conducted, and will be discussed in more detail in the chapter on economizers (LBNL 2007b).

Energy Loss Due to High Filter Pressure Drop and/or Poor Filter Maintenance. The difference in the design pressure drop of a clean vs. dirty filter is often a factor of two or more. There may also be a factor of two difference on the initial (clean) pressure drop of two competing filters with the same filtration efficiency. A TCO analysis is needed to determine the most cost-effective initial pressure drop for a filter, and also the frequency of maintenance/replacement. Often the pressure drop across filters is monitored to ensure that it does not get too dirty, as this can adversely affect both energy consumption and overall system performance.

Savings Example #2: Filtration Cost Savings

Assume that a 10,000 cfm outdoor air system has a filtration system that requires 1.50 in. w.g. pressure drop. What is the annual fan energy cost savings if a filter is provided that can provide the same filtration at 0.75 in. w.g. pressure drop? Assume continuous operation and fan energy costs of $0.10/kWh.

$$\text{Brake Hp} = \frac{0.75 \text{ in. w.g.} \times 10,000 \text{ ft}^3/\text{min}}{(6356 \times \text{motor eff.} \times \text{fan eff.})}$$

Assuming a fan efficiency of 70%, and a motor efficiency of 93%, the brake Hp savings in this example would be the following:

$0.75 \times 1000 / (6356 \times 0.7 \times 0.93) = 1.81 \text{ Hp} \times 0.746 \text{ kW/Hp} = 1.35 \text{ kW}$

Annual savings = 1.35 kW × 8760 h/yr × $0.10/kWh = $1183/yr

2.3.4 Ventilation

Outside air is introduced to the data center space for one of the following four reasons: to maintain indoor air quality requirements, to pressurize the space to keep contaminants out, as make-up air for smoke purge, or to conserve energy when outside air conditions are conducive to provide free cooling.

Ventilation air should be preconditioned to a dew-point temperature within the operating range of the datacom space. Indoor air quality requirements are spelled out in *ANSI/ASHRAE Standard 62.1-2004, Ventilation for Acceptable Indoor Air Quality* (ASHRAE 2004k), or the latest version of this standard. Although not inclusive of all the air quality requirements for datacom environments, it helps to ensure that adequate quantity and quality of ventilation air is available to all building occupants.

The major energy impacts of ventilation to a datacom facility are:

- Outdoor air-conditioning costs
- Dew-point control of incoming outdoor air

Outdoor Air-Conditioning Costs. While for many hours of the year outdoor air is more costly to condition to datacom requirements than return air from the facility, there are certain times of the year when outdoor air can be brought into the space to provide "free" cooling. This type of system is called an air-side economizer, and it can substantially reduce the operating cost of the air-conditioning system for (typically) many thousands of hours per year. Chapter 4 of this publication discusses air-side economizer cycles in more detail. The use of an air-side economizer must take into account the outside contamination levels and the possibility of datacom equipment damage without proper filtration of particulates as well as gas contaminants.

Dew-Point Control of Incoming Outdoor Air. The only significant source of humidity in most datacom facilities is the outdoor air. Due to the control difficulties (i.e., fighting) that have occurred in some datacom facilities in trying to maintain relative humidity at the zone level, many datacom facility designers have recognized that if the dew point of the entering air can be controlled to provide a relative humidity in the proper range for the facility's datacom equipment, there may not be a need for any humidification or dehumidification at the zone (or CRAC unit) level. Alternately, CRAC unit controls can be set to control absolute humidity, and/or units can incorporate a unit-to-unit communication scheme which allows them to coordinate as a team.

2.3.5 Envelope Considerations

There are several parameters that should be considered in designing the envelope of datacom facilities for energy-efficient operation. However, energy-efficient envelope design for data centers is not much different than for other facilities. As such, it will not be covered further here, and the reader is referred to the *ASHRAE*

Handbook—Fundamentals (ASHRAE 2005b) and ASHRAE Standard 90.1 (ASHRAE 2004a) for further guidance.

2.4   ENERGY-EFFICIENCY RECOMMENDATIONS/BEST PRACTICES

Environmental conditions can have a substantial impact on energy efficiency and TCO in a datacom facility. Many energy-efficient concepts have been presented in this chapter. Those likely to have the greatest impact on datacom facility energy consumption are as follows:

- Operating in the range recommended by ASHRAE's thermal guidelines can result in significant energy savings due to greater refrigeration cycle thermodynamic efficiency and increased economizer hours.
- Allowing for increased temperature and humidity dead bands will eliminate "fighting" between adjacent supply air units, which is a significant source of inefficiency in some existing facilities. Use of reheat coils in CRAC units, or other means of simultaneous heating and cooling, should concurrently be eliminated.
- Dew-point control of a data center is an efficient means of controlling both humidification and dehumidification.
- Filters should be sized for low initial pressure drop, and should be maintained on a regular basis to keep filtration energy losses low.

3

Mechanical Equipment and Systems

3.1 INTRODUCTION

The mechanical systems that provide for environmental control of temperature and humidity in a data center are significant users of energy. As shown in Figure 3.1, case studies of 12 data centers by LBNL found that the cooling plant accounted for an average of 23% of the total energy consumption of the facility, and the air-moving equipment accounted for an average of 8% (LBNL 2007a). In total, HVAC systems thus account for about 31% of facility energy consumption. While most facilities had

Average Data Center Power Allocation

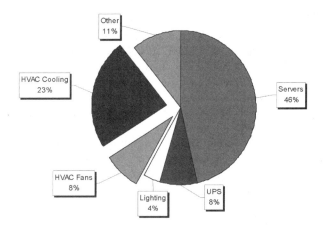

Figure 3.1 Energy consumption impact of mechanical equipment and systems (LBNL average of 12 data centers).

HVAC system energy consumption near the 31% average, the range for this usage was 20% for the most efficient system and 54% for the least efficient system, indicating that close attention needs to be given to HVAC systems to minimize datacom facility energy consumption and total cost of ownership (TCO).

The purpose of this chapter is to discuss the typical components that make up the mechanical support systems in a datacom facility, and to provide guidance on minimizing the energy consumption of these components. In most cases the components need to be considered as part of overall systems to optimize energy efficiency.

This chapter is divided into three broad equipment categories. The three categories are cooling distribution equipment, mechanical cooling equipment, and heat rejection equipment. For reference, two diagrams showing typical HVAC system layouts for a datacom facility, with these three categories, are provided as Figures 3.2a and 3.2b. Figure 3.2a shows a computer room air handler (CRAH) unit with a chilled-water cooling coil. Chilled-water cooling is typical in large facilities. Figure 3.2b shows a computer room air conditioner (CRAC) unit that is more typically found in smaller facilities. Figure 3.2a also lists a number of possible alternatives for cooling distribution equipment, mechanical cooling equipment, and heat rejection equipment.

3.2 COOLING DISTRIBUTION EQUIPMENT

The mechanical equipment that tends to be most specialized to datacom facilities is the cooling distribution equipment. The efficient air distribution of this equipment is covered in Chapter 5 and will not be covered in this section. What will be covered here are the design considerations of this equipment that affect component energy consumption.

3.2.1 Fans

Fan energy consumption in distribution equipment can be significant, especially since most fans in a datacom facility run 24 hours per day, 365 days per year. Energy efficiency can be optimized in many ways, including the following:

- Design the air distribution (supply and return) systems with minimal pressure drop—this typically means decreasing pathway distances and/or decreasing restrictions to flow
- Maintain a high temperature differential across heat exchangers associated with fans to minimize fan "pumping power" per unit of heat transfer
- Utilize a fan that has high efficiency at the expected operating condition(s)
- Utilize premium-efficiency motors
- Utilize low-pressure drop filters
- Provide frequent filter maintenance
- Utilize variable-frequency drives (VFD) or variable-speed electronically commutated direct-drive fans where applicable
- Adopt efficient airflow distribution techniques (refer to Chapter 5)

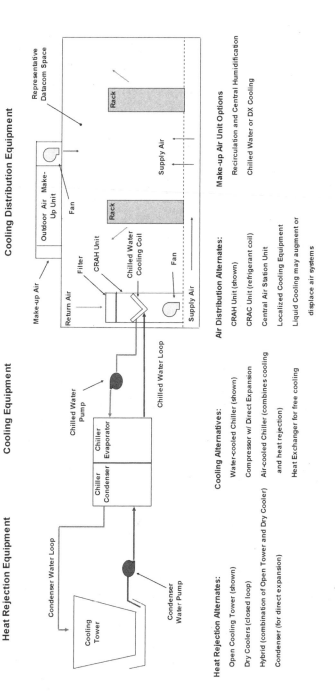

Figure 3.2a Datacom facility HVAC schematic using CRAH units, with list of alternatives.

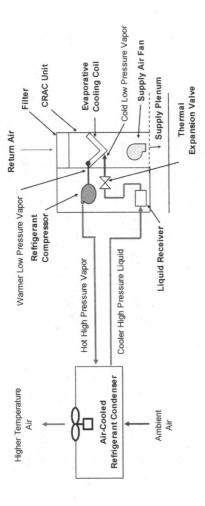

Figure 3.2b Datacom facility HVAC schematic using CRAC units with air-cooled condenser for heat rejection.

A properly designed variable air volume (VAV) system may be the best approach to maintaining fan energy consumption proportional to datacom equipment energy use. Even if the decision is made not to utilize an active VAV system, a variable-speed drive can still be used during the commissioning process to more closely match fan speed to the required airflow of the facility to maintain high efficiency. A variable-speed drive can also be used to maintain constant airflow by adjusting fan frequency to counterbalance varying pressure drop from filter loading.

3.2.2   Chilled-Water Pumps

Chilled-water pumps typically operate 24 hours per day, 365 days per year, so close attention should be paid to pump selection. Attention to the following items should result in optimal pump energy efficiency.

*   The pump selection should be optimized for the operating point. Several manufacturers and pump models should be checked to find the most efficient pump for the application.
*   The ΔT of the chilled water should be optimized. A greater ΔT will result in lower flow and, thus, lower pumping energy consumption, but potentially lower chiller efficiency or larger chilled-water coils.
*   Premium-efficiency motors should be selected.
*   As long as all system components are designed for variable-speed operation, variable-speed drives should be specified. Various piping schemes should be considered, including variable-speed primary pumping (typically considered the most efficient), and primary/secondary pumping systems.

More information on efficient pumping system design can be found in the "Hydronic Heating and Cooling System Design" chapter of the *2004 ASHRAE Handbook—HVAC Systems and Equipment* (ASHRAE 2004g).

3.2.3   Cooling Coils

There are several parameters of a cooling coil to be considered in optimizing energy efficiency, including (a) water pressure drop, (b) air pressure drop, (c) entering and leaving air-side conditions, and (d) entering and leaving water-side conditions. All four of these parameters interact with each other in coil design, and an algorithm to minimize energy while still achieving proper performance needs to be developed in each case. CRAC unit manufacturers typically provide (a) and (b) as a function of (c) and (d), so these data are readily available. Since the datacom load is primarily sensible, the cooling coil should be selected with a high sensible heat ratio.

Figure 3.3 shows the trade-off between air pressure drop and approach temperature for a sample chilled-water coil, where

Approach temperature = air temperature leaving coil
– water temperature entering the coil.

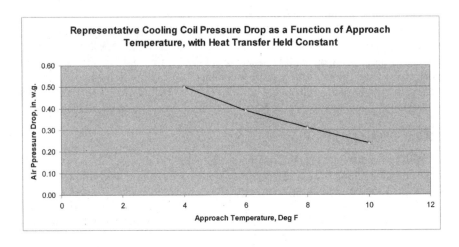

Figure 3.3 Representative cooling coil air pressure drop as a function of approach temperature between entering chilled-water temperature and leaving air temperature.

Figure 3.3 demonstrates that the coil pressure drop (and, thus, fan energy) decreases as the approach temperature increases for a given heat transfer capacity. With just this component in mind, a decision would be made to design a coil for a high approach temperature. Unfortunately, a higher approach temperature will typically decrease chiller efficiency and economizer hours, so an analysis that can combine all parameters on an annualized basis is needed for optimization.

3.2.4 Humidification Equipment

There are a number of ways to introduce humidity into a data center to achieve the desired humidity range. The choice of which humidification technology to use is typically based on the following three design characteristics:

1. Water quality (conductivity)
2. System ease of maintenance
3. Energy efficiency

The technology with the widest water quality tolerance is the infrared system, as it has no conductivity requirements. Steam generation is the simplest to maintain via replacement of canister bottles, but this may not be the least expensive depending on the frequency of exchange required (dependent of the mineral content of the water). The most energy efficient is the ultrasonic, provided a supply of de-ionized water is available. In addition to its low operating cost, ultrasonic humidifiers will help satisfy the cooling load due to its adiabatic cooling process, which provides approximately 1000 Btuh (0.3 kWh) of free cooling for every 1 lb/h (0.45 kg/h) of humidification. Note, the cost to generate de-ionized water must be considered in the TCO analysis.

The humidifier must be responsive to control, maintainable, and free of moisture carryover. The humidity sensor should be located to provide control of air inlet conditions to the IT equipment.

When humidification is provided from a central system, additional options include adiabatic wetted media, compressed air fogging units, and natural gas-fired steam systems. The unique environmental conditions of datacom facilities often lend themselves to utilizing a humidification system that is independent from the primary cooling system. The reason is that, particularly for underfloor supply, the discharge temperature of the air handlers is often the same temperature as the inlet to the datacom equipment. If the humidifier is installed after the cooling coil, the design needs to incorporate bypass air to facilitate proper absorption of moisture.

Additional reference information on humidifiers can be found in the "Humidifier" chapter of the *2004 ASHRAE Handbook—HVAC Systems and Equipment* (ASHRAE 2004d).

3.2.5   Dehumidification and Reheat Coils

Most dehumidification equipment used in data centers is refrigeration-based, i.e., based on the use of a cooling coil to dehumidify the airstream and condense a certain percentage of the moisture. In many cases, the cooling coil utilized for temperature control is the same as that used for dehumidification.

Reheat coils are available as an option from most (if not all) CRAC unit manufacturers as a means to maintain the required supply temperature when overcooling occurs as a result of trying to remove excess moisture form the air via the cooling coil. Since the recommended relative humidity at the inlet to datacom equipment (per ASHRAE [2009]) is 41.9°F dew point–60% RH and 59°F dew point, reheat coils have historically been used to increase the supply temperature (and reduce the relative humidity) of the supply air to the datacom equipment.

One of the ways to minimize the energy consumption of reheat coils is to control the room moisture level based on absolute humidity, not relative humidity.

Another way to minimize energy consumption of reheat coils is to simply eliminate them. Many data centers have an independent humidity control system, such as a low-temperature central system cooling coil, to control dew point in the data center. This allows the cooling coils in the CRAC to modulate solely to control temperature, and the need to overcool and reheat for humidity control is eliminated. If system designers determine that a reheat coil is required for proper environmental control, consider the use of waste heat for the reheat coil, if a source is available. Hot gas reheat from the compressors is available in packaged DX humidity control units and DX-cooled CRAC units.

3.2.6   CRAH/CRAC Unit Energy Considerations

CRAH and CRAC units are popular for conditioning and control of datacom facilities. Rating of CRAH/CRAC units, including energy-efficiency ratings, are covered by ANSI/ASHRAE Standard 127-2007 (or latest version) (ASHRAE 2007i). The 2007 revision took many significant steps toward defining an efficiency measurement approach that more closely matches an applied efficiency and helps to expose the lowered energy consumption available due to various best practices. Some of the attributes now covered by the standard include the following:

1. The efficiency metric is based on the sensible capacity of the unit, not the total capacity.
2. The system static pressure is now based on more realistic static pressures that vary based on the type of airflow pattern (downflow, upflow to ducting, upflow to open plenum).
3. The utilization of variable capacity compressors now has an impact on the measurement of a unit's efficiency.

A diagram showing the basic components of typical CRAH and CRAC units is shown in Figure 3.4. As this diagram shows, the basic difference between these units is that the CRAH unit utilizes a chilled-water cooling coil, while the CRAC unit utilizes a refrigerant with an evaporative cooling coil and integral compressor.

Energy-consuming components in CRAH/CRAC units may include any of the following:

*   Compressors
*   Fan systems
*   Cooling coils (remote energy source)
*   Reheat components (can be electric, hot water, hot gas)
*   Humidifiers (several types are available)
*   Heat rejection devices—usually a dedicated remote condenser, dry cooler, or wet cooler.

In addition to the observations already made in this chapter, the following should also be considered in CRAH/CRAC system design for energy efficiency:

*   The thermodynamic efficiency of a central plant (especially a water-cooled plant with centrifugal chillers) is usually higher than the efficiency of a compressor found in the typical 10–60 ton CRAC unit. (This is a significant factor in the trend toward central cooling plants for large datacom facilities.)
*   CRAH/CRAC unit reheat coils are almost never required in an efficiently designed system.
*   Integral fluid economizers are typically easier to deploy on modular CRAC units, while integral air economizers are easier to deploy on central air handlers.

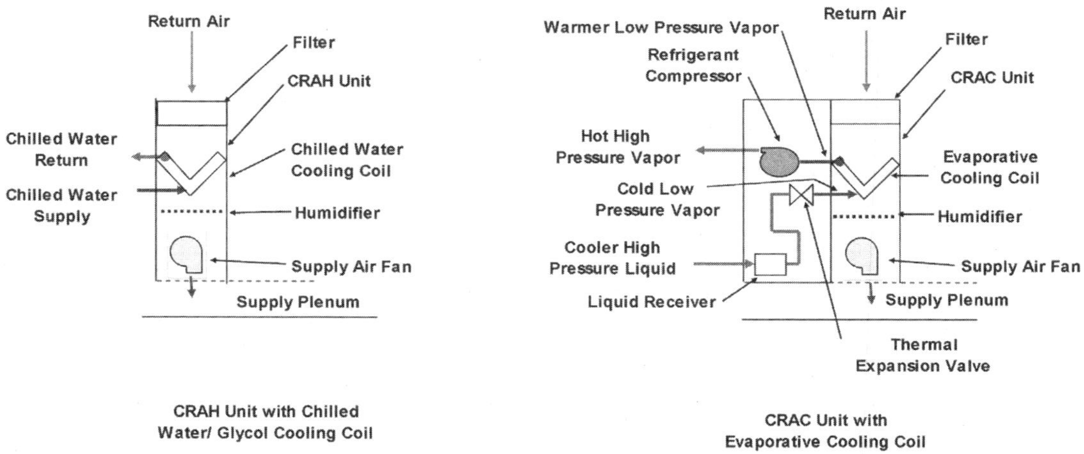

Figure 3.4 Comparison of CRAH and CRAC units.

3.3   MECHANICAL COOLING EQUIPMENT

Cooling equipment comprises the central portion of the system diagram in Figure 3.2. The two types of mechanical cooling equipment typically used to cool data centers are (1) water chillers and (2) small refrigerant compressors. The choice of equipment can be related to a number of factors, but the predominant factor is size, with compressorized units used for small facilities, and chilled-water systems used for larger facilities.

Chillers are typically one of the largest users of energy in the larger datacom facilities. As such, efforts to maximize the energy efficiency of a chiller, or to minimize hours of operation, can go a long way in terms of reducing TCO. Much of the information on the next few pages is well known to those experienced in chiller design, but it is included here to help those in the datacom industry understand the parameters that affect chiller efficiency. Parameters that will be discussed here include:

- Type and size of chiller
- Chilled-water supply temperature
- Chilled-water differential temperature
- Entering condenser-water temperature
- Condenser-water differential temperature
- Part-load efficiency as a function of compressor VFD drive

3.3.1   Types and Sizes of Chillers

There are many different types of chillers, and designs are covered well in existing publications, such as in the "Liquid-Chilling Systems" chapter of the *2004 ASHRAE Handbook—HVAC Systems and Equipment* (ASHRAE 2004l). Types of chillers include:

- Reciprocating
- Scroll
- Screw
- Centrifugal
- Absorption

In general, the reciprocating and scroll chillers are relatively small in size (less than 100 tons), while the screw and centrifugal compressors are generally used for larger machines. As would be expected, the larger chillers tend to be more efficient than the smaller ones.

Absorption chillers are generally considered to be a separate class of chiller. Rather than having an electric-motor-driven compressor, they use a heat source. Their COP is typically much less than a compressor-driven chiller, but they are popular where there is a free or low-cost source of heat. For more information, refer to the "Absorption Cooling, Heating, and Refrigeration Equipment" chapter of the *2006 ASHRAE Handbook—Refrigeration* (ASHRAE 2006b).

Another chiller variable is refrigerant type; there are many options available, with different safety, ozone depletion, and global warming impacts if the refrigerant escapes to the environment. A good source of comparative information for safety is *ANSI/ASHRAE Standard 34-2007, Designation and Safety Classification of Refrigerants* (ASHRAE 2007a). For ozone depletion and global warming, a good source is an article by Calm and Didion (1997). The USEPA Web site also has information on the ozone depletion potential (ODP) and global warming potential (GWP) of refrigerants (USEPA 2007). Some chiller/refrigerant combinations have peak efficiency ranges, i.e., sweet spots for certain temperatures and/or temperature differentials. It is a good idea to examine the expected full-load and part-load design efficiency of chillers and refrigerants from several vendors to determine the best choice for a project.

3.3.2 Chilled-Water Supply Temperature

The thermodynamic efficiency of a chiller is sensitive to the difference in temperature between the chilled water and the condenser water. Chiller efficiency can increase dramatically if this differential temperature is lowered. Figures 3.5a and 3.5b show, for a sample chiller, the impact of chilled-water supply temperature on chiller efficiency, for a constant chilled-water differential temperature and condenser-water temperature. Figure 3.5a shows this efficiency in units of kW/ton, where "kW" refers to the electrical input kW to the chiller and "ton" refers to the cooling output of the chiller (1 ton = 12,000 Btu/h [3.52 kW]). Figure 3.5b shows this same relationship measured as a coefficient of performance (COP).

Many facilities have historically used 45°F (7.2°C) chilled water as the supply temperature to a facility. Combined with a 7°F (3.9°C) cooling coil approach temperature, 52°F (11.1°C) supply air near saturation could be supplied to a facility, resulting in a space condition of 72°F (22.2°C) and 50% relative humidity. For data centers, chilled-water temperature temperatures higher than 45°F (7.2°C) should be considered to take advantage of the efficiency gains shown in Figure 3.5. Datacom equipment air inlet temperatures as high as 80°F (27°C) are still within the recommended operating range of almost all datacom equipment, and since internal sources of humidity are low, small low-temperature coils in the facility (possibly in the make-up air units) can probably handle the dehumidification load of the facility, allowing the primary cooling coils (whether in CRACs or in central air handlers) to operate dry at a higher temperature.

3.3.3 Chilled-Water Differential Temperature

Figure 3.6 shows, for an actual chiller, the relationship between chiller efficiency and the differential temperature of the chilled water across the evaporator. In

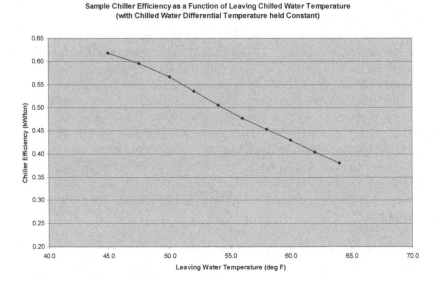

Figure 3.5a Sample chiller efficiency as a function of leaving chilled-water temperature, units of kW/ton (with all other parameters held essentially constant).

this case, the chiller efficiency is quite flat as a function of differential temperature. From an energy-efficiency standpoint, therefore, it is probably advantageous to increase this differential temperature, as pumping costs will be reduced with little or no effect on chiller energy consumption. It should be noted that the data for Figure 3.6 comes from a sample chiller for the purposes of demonstrating a point—the system designer needs to obtain product-specific curves for their equipment.

3.3.4   Condenser-Water Differential Temperature

Figure 3.7 shows, for an actual chiller, chiller efficiency as a function of the condenser-water differential temperature. In this case, there is clearly a decrease in chiller efficiency as the condenser-water differential temperature increases. Increasing the differential temperature to achieve reduced pump energy will clearly result in a less-efficient chiller. A more detailed analysis of condenser-water pump energy and chiller energy, on an annualized basis, is required to determine the optimized operating point.

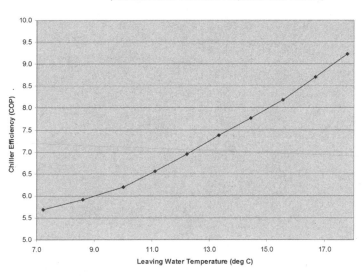

Sample Chiller Efficiency as a Function of Leaving Chilled Water Temperature
(with Chilled Water Differential Temperature held Constant)

Figure 3.5b Sample chiller efficiency as a function of leaving chilled-water temperature, units of COP (with all other parameters held essentially constant).

3.3.5 Variable-Speed Compressor Drives

Chillers with variable-speed compressors have been available for more than 20 years, and are a well-established method to achieve high energy efficiency at part-load operation. Figure 3.8 compares the part-load performance of a chiller with constant speed operation versus one with variable-speed operation. The capital cost of a chiller with a VFD drive is greater, but the figure shows the significant increase in part-load efficiency possible with the VFD drive.

Another strategy for maintaining good part-load performance is to have multiple chillers, staged as needed to match the datacom facility cooling load.

3.4 HEAT REJECTION DEVICES

Heat rejection to the exterior of the building typically occurs to the atmosphere, though rejection to bodies of water, or to the ground, may also be possible for specific locations. The most common forms of heat rejection in the datacom industry are cooling towers and dry coolers, which are described in more detail below. For the location of the heat rejection devices as part of the larger cooling and heat rejection system, refer to Figure 3.2.

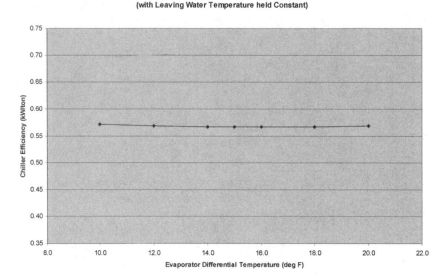

Figure 3.6 Sample chiller efficiency as a function of evaporator differential temperature (with other parameters held essentially constant).

3.4.1   Open Cooling Towers

The definition of a cooling tower is "a heat-transfer device, often tower-like, in which atmospheric air cools warm water, generally by direct contact (evaporation)" (ASHRAE 1991). Cooling towers are an efficient means of heat rejection. There are several types of cooling towers, and designers are encouraged to contact vendors to determine the most efficient tower for a specific climate and application. Figure 3.9 shows the most common type of cooling tower, an open tower where the condenser water is cooled by direct contact with the ambient air.

Figure 3.10 shows, for a sample chilled-water system, the relationship between cooling tower approach temperature and chiller efficiency. The figure shows that the chiller efficiency increases as the tower approach temperature decreases, due to lower condenser-water supply temperature to the chiller. In this case, the increase in chiller efficiency needs to be weighed along with the increase in cooling tower capital cost required to achieve this closer cooling tower approach, and along with the probable increase in cooling tower fan energy needed to achieve the same result. In many cases, an "oversized" cooling tower can be used to achieve the same heat rejection as a smaller tower, but with less fan energy consumption.

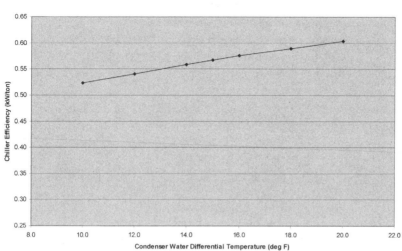

Figure 3.7   Sample chiller efficiency as a function of condenser-water differential temperature (with other parameters held essentially constant).

In addition to designing a tower for a low approach temperature to increase chiller efficiency, strategies to limit the fan energy of the chiller must also be considered. Fan efficiency considerations include the following:

- Tower fan energy consumption can be quite variable for a given cooling load. Check several vendors, and several tower models from each vendor, to optimize tower selection for efficient operation.
- Choose the best fan for the job. Propeller fans typically have lower unit energy consumption than centrifugal fans.
- Towers designed for a high water flow turn-down ratio can allow for condenser-water flow better matched to chiller needs, especially in plants with multiple chillers. The term *turn-down ratio*, used in this context, is the ratio of the highest to the lowest recommended water flow for the cooling tower.
- Premium-efficiency motors and variable-speed drives should be specified.

Maintenance and make-up water are issues that must be addressed with evaporative towers, and water conditioning to prevent scaling and bacterial and/or algal growth is essential. More information on cooling towers can be obtained in the

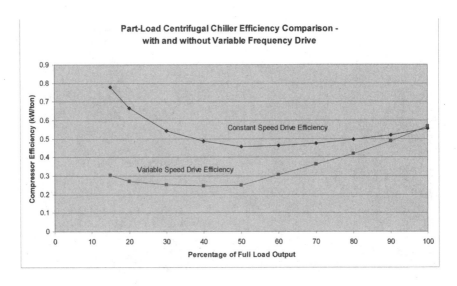

Figure 3.8 Sample part-load centrifugal chiller efficiency with and without VFD.

"Cooling Towers" chapter of the *2004 ASHRAE Handbook—HVAC Systems and Equipment* (ASHRAE 2004f).

3.4.2 Dry Coolers

A dry cooler, in its simplest form, is a heat-transfer device that utilizes atmospheric air blowing over a closed-loop water or glycol coil to transfer heat from the liquid to the air. Figure 3.11 shows a schematic of a dry cooler.

Evaporative heat rejection systems, such as cooling towers, are almost always more efficient than dry systems because (1) wet-bulb temperatures are always at or below dry-bulb temperatures and (2) wetted surfaces, or surfaces with evaporation taking place, have greater heat-transfer coefficients and, thus, closer approach temperatures than dry coils. There are instances, however, where dry heat rejection is the preferred method. Obstacles in relation to water-cooled heat rejection may include restricted water availability (or sewer availability for blowdown), concerns about legionellosis, and/or concerns about freezing with open tower systems. Small heat rejection systems, such as those serving an individual CRAC unit, are also often air cooled, as the cooling capacity of a CRAC unit is usually smaller than the heat rejection capacity of a small cooling tower.

The only energy use specific to the dry cooler equipment is fan energy. The primary system energy trade-off is that closer approach temperatures (which will

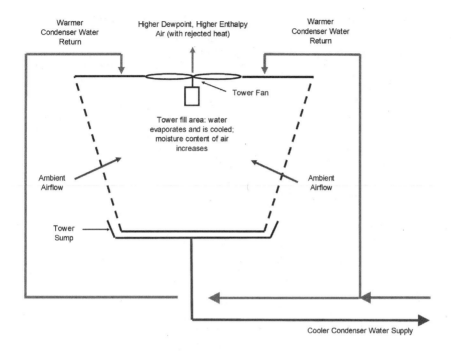

Figure 3.9 Open cooling tower schematic.

improve chiller or compressor efficiency) come at the cost of higher airflow velocities or deeper coils, which require higher dry cooler fan energy.

Considerations in the optimization of dry cooler efficiency include:

- Larger coil-to-fan size ratios will increase energy efficiency (and typically reduce noise)
- Efficient fans
- Efficient fan motors
- Efficient part-load control of the fans
- Some "dry coolers" run dry for 90% of the year, but have evaporative spray cooling for the remaining 10% of the year to achieve a lower leaving water temperature on the hottest days of the summer (resulting in considerable electrical demand and energy savings)
- Use of dry coolers for water-side economizer cooling (typically for CRAC units)

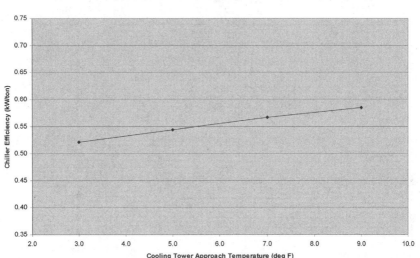

Sample Chiller Efficiency as a Function of Cooling Tower Approach Temperature
(with Condenser Water Differential Temperature held Constant)

Figure 3.10 Sample chiller efficiency as a function of cooling tower approach temperature (with all other parameters held essentially constant).

The last item, a water-side economizer, is a feature for most CRAC unit manufacturers, and is illustrated in Figure 3.12.

If the temperature of the water leaving the dry cooler is less than the temperature of the return air to the CRAC unit, this water can be used to precool the air entering the primary CRAC unit cooling coil, thus reducing the load on this coil. The fan and pump energy increase in the system must be compared to reduced compressor energy use, which is system and climate specific. The efficiency gains are greatest if the compressors are specifically designed to take advantage of lower condenser fluid temperatures when the economizer coil is working at partial load.

3.4.3   Hybrid and Other Systems

Some ambient heat rejection systems are a combination of wet and dry cooling. One such system is the evaporative-cooled chiller, which is an air-cooled chiller with a spray coil. There are also hybrid cooling towers, and, as indicated above, wet/dry coolers that utilize spray cooling for some hours in the year. Hybrid heat rejection systems will not be covered in any detail in this book, but designers should investigate these options to find the most efficient operating system, particularly if an open-

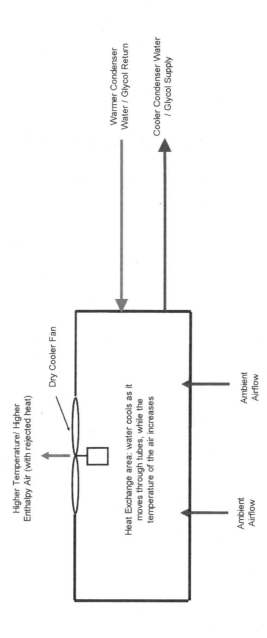

Warmer Condenser Water / Glycol Return

Cooler Condenser Water / Glycol Supply

Dry Cooler Fan

Higher Temperature/ Higher Enthalpy Air (with rejected heat)

Heat Exchange area: water cools as it moves through tubes, while the temperature of the air increases

Ambient Airflow

Ambient Airflow

Figure 3.11 Dry cooler schematic.

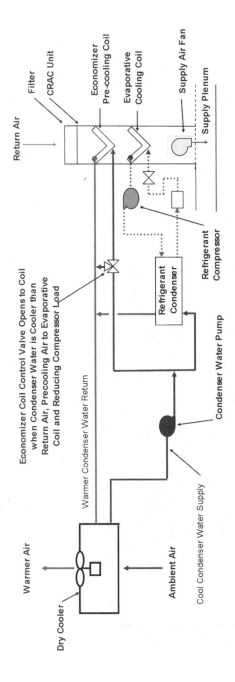

Figure 3.12 CRAC unit schematic with economizer precooling coil.

loop evaporative cooling tower system is not possible. More information on hybrid cooling towers can be obtained in the "Cooling Towers" chapter of the *2004 ASHRAE Handbook—HVAC Systems and Equipment* (ASHRAE 2004f).

3.4.4   Condenser-Water Pumps

Some condenser pumps may only be operational when the chillers are operational. Others may be operational for the whole year (similar to chilled-water pumps), especially if they serve year-round cooling towers.

There are several considerations in condenser-water pump selection that will impact energy efficiency and TCO.

*   The pump selection should be optimized for the operating point. Several manufacturers and pump models should be checked to find the most efficient pump for the application.
*   The ΔT of the condenser water should be optimized. A greater ΔT will result in lower pumping costs, but potentially lower chiller efficiency.
*   Premium-efficiency motors and, where applicable, variable-speed drives should be specified.
*   Modern chillers can provide control signals to directly or indirectly control condenser-water flow and temperature.
*   The pumping system should be designed to minimize energy consumption: if one portion of a chilled-water system serves a condenser that is only used for part of a year, for instance, this should be on a separate loop than primary tower water that may be required year round.

3.4.5   Heat Exchangers

Heat exchangers are often an integral part of the heat rejection system in a data center, and are used during low ambient conditions to achieve free cooling. The most important design characteristics of heat exchangers are the approach temperature and the pressure drop. The trade-off, which may need computer analysis to optimize on an annualized basis, is that a closer approach temperature (which will increase economizer hours of operation) can be obtained with higher flow velocities, but this comes at the expense of increased pump energy.

More information on the systems in which heat exchangers are used in datacom facilities is provided in Chapter 4, "Economizer Cycles." Theoretical information can also be found in the "Heat Exchanger" chapter of the *2004 ASHRAE Handbook—HVAC Systems and Equipment* (ASHRAE 2004).

3.4.6   Refrigerant Condensers

As indicated previously, CRAC units with integral refrigerant compressors are typically used in the smaller datacom facilities, where the heat load does not justify a complete chilled-water plant (or where other considerations preclude the use of chilled water or glycol). A schematic diagram of a CRAC unit, with a liquid-cooled refrigerant condenser for heat rejection, has already been shown in Figure 3.12. A schematic diagram of a CRAC unit, with an air-cooled refrigerant condenser for heat rejection, is shown in Figure 3.13. More specific information on refrigeration cycles for CRAC units can be obtained from the various manufacturers of this equipment.

3.5 SYSTEM AND EQUIPMENT DESIGN FOR EFFICIENT PART-LOAD OPERATION

As described in Chapter 1, technology is continually changing and, therefore, datacom equipment in a given space is frequently changed and/or rearranged during the life of a datacom facility. As a result, the HVAC system serving the facility must be sufficiently flexible to permit "plug and play" rearrangement of components and expansion without excessive disruption of the production environment. To guarantee a high-efficiency cooling system in the face of uncertain cooling loads, systems should be designed to effectively and efficiently handle a wide range of heat loads. Computer simulation programs, or other means to ensure that high efficiency is maintained through the range of expected climatic conditions and part-load operation, are highly recommended.

3.6 ENERGY-EFFICIENCY RECOMMENDATIONS/BEST PRACTICES

As this chapter has hopefully demonstrated, there are many components to the energy-efficient design of the mechanical cooling system equipment, and associated distribution and heat rejection systems. The following points highlight some of the design parameters that are common to many components in these systems, or to the cooling system as a whole:

- For each component of the system, examine several alternatives—the first selection may not be the most efficient. Communicate with consultants, vendors, maintenance personnel, and others with experience and expertise to determine the best component for a specific project and cooling system.
- Design the system and choose components that operate efficiently at part-load as well as at full-load. The load in a datacom facility is not fixed over time, so flexibility in this regard is paramount.
- The use of VFDs (or equivalent means of achieving reduced energy consumption at part-load operation) will almost certainly yield a good payback, as well as conserve energy.
- Chillers are a very significant component of overall energy consumption in water-cooled facilities; an understanding of the parameters that most affect chiller efficiency is paramount to designing and operating an efficient mechanical cooling plant.

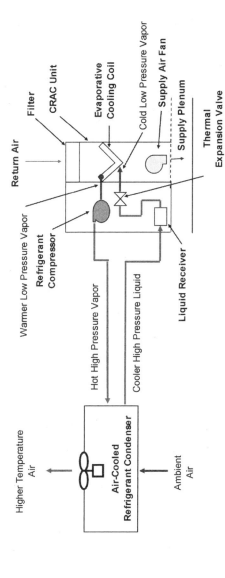

Figure 3.13 CRAC unit schematic with air-cooled condenser for heat rejection.

- For CRAC units, be aware that the efficiency metrics in the revised ASHRAE Standard 127 are now more focused on sensible cooling capacity rather than total cooling capacity
- The energy usage of the mechanical cooling plant system should be modeled over an entire calendar year, and with various facility loads, to determine the most efficient combination of components for both full-load and part-load operation. Examine many options—finding the best option in the modeling phase is considerably more efficient than making a change later in the design or construction process.

4

Economizer Cycles

4.1   INTRODUCTION

A short definition for an *economizer* is "a control system that reduces the mechanical heating and cooling requirement" (ASHRAE 1991). The standard meaning of this definition in the HVAC industry is the utilization of outdoor air under certain conditions to allow chillers and/or other mechanical cooling systems to be shut off or operated at reduced capacity. Operating in an economizer mode is often also known as "free cooling." There are a number of different ways to achieve free cooling, and applicability to a specific project is a function of climate, codes, performance, and preference. Data centers differ from office environments in (1) the need for humidification control, (2) concerns about contamination from outdoor air sources adversely affecting datacom equipment performance, and (3) the potential—in some centers, and after careful analysis—to provide higher supply air temperatures, which can, in turn, increase economizer cycle hours on an annual basis. The impact of economizers is almost exclusively confined to the "HVAC Cooling" slice of the energy pie shown in Figure 4.1. Case studies of 12 data centers by LBNL found that the cooling plant accounted for an average of 23% of the total energy consumption of a datacom facility (LBNL 2007a). This section will start with a brief code overview of economizers, and then discusses three broad categories of economizer systems: air-side economizers, adiabatic-cooled air-side economizers, and water-side economizer systems. A final section will compare the climatic advantages of each category of economizers.

4.2   ECONOMIZER CODE CONSIDERATIONS

ASHRAE Standard 90.1 discusses requirements for both air-side and water-side economizers (ASHRAE 2004a). For those following the prescriptive path of Standard 90.1, Section 6.5.1, "Economizers," states that "each cooling system having a fan shall include either an air or water economizer meeting the requirements of 6.5.1.1 through 6.5.1.4." The headings for these four paragraphs are "Air Econ-

Average Data Center Power Allocation

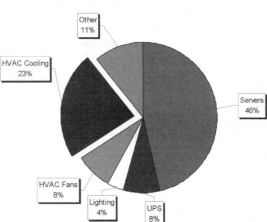

Figure 4.1   Energy consumption impact of economizers (LBNL average of 12 data centers).

omizers," "Water Economizers," "Integrated Economizer Control," and "Economizer Heating System Impact," respectively.

ASHRAE Standard 90.1 has several exemptions to the economizer requirement, including an exemption for facilities where "more than 25% of the air designed to be supplied by the system is to spaces that are designed to be humidified above 35°F (2°C) dew-point temperature to satisfy process needs." Class 1 and Class 2 data centers designed to comply with *Thermal Guidelines for Data Processing Environments* (ASHRAE 2009) fall into this exemption category. Still, it is fair to say that many datacom facilities will find that their TCO is reduced if an economizer, either air or water, is installed. The economics of economizers typically look better with larger facilities.

Further information on energy codes as they relate to data centers can be found in Appendix D. All applicable energy codes should be referenced for a specific locale, as state and local codes may differ from the ASHRAE standard, and adoption dates also vary from state to state.

4.3   AIR-SIDE ECONOMIZERS

An air-side economizer is a system that has been designed to use outdoor air for cooling—partially or fully—when the outdoor air meets certain criteria. Air dampers are used to adjust the mixture of outdoor air and return air. Figure 4.2a shows a

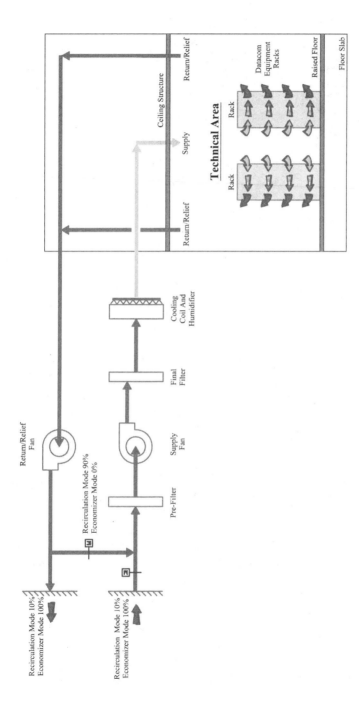

Figure 4.2a Air-handling system with air-side economizer (see Figures 4.3a and 4.3b for process).

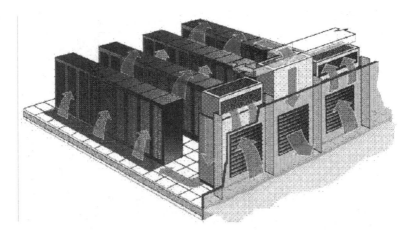

Figure 4.2b Configuration of a CRAC with air-side economizer.

diagram of an air handler with a mixing box and return/relief air fan that provides for the necessary redirecting of airstreams for an economizer cycle. Air-side economizers are typically better suited for central station air handlers than CRAC units that are often used for computer room cooling because the central station air handlers can be located close to outdoor air sources such as exterior walls and roofs. However, air-side economizers for CRAC units are available, as depicted in Figure 4.2b.

While not shown in the diagrams, some economizers use dry-bulb temperature sensors to control the damper positions, whereas enthalpy economizers use a combination of temperature and humidity sensors to determine the total heat content of the air, and make the switchover decision to minimize the total (as opposed to strictly sensible) cooling load. Enthalpy measurement of both outdoor air and return air is recommended for datacom facilities, and this recommendation applies to both the dry air economizer and the adiabatic-cooled air economizer. Care must be taken, however, to calculate the energy costs of dehumidifying outdoor air, as will be discussed in more detail shortly.

Figure 4.3a shows examples of the conditions associated with an air-side economizer on a psychrometric chart. Figure 4.3b, in turn, shows ambient environmental regions corresponding to zones where different controls and/or environmental conditioning are required to optimize energy efficiency. The two examples of psychrometric processes shown in Figure 4.3a (one "winter" and one "spring") correspond to ambient conditions in Regions I and II of Figure 4.3b.

Region I of Figure 4.3b corresponds to typical winter ambient conditions where outdoor air, varying in quantity between the minimum and 100%, is blended with the return air from the data center as required to achieve the desired supply air dry-bulb

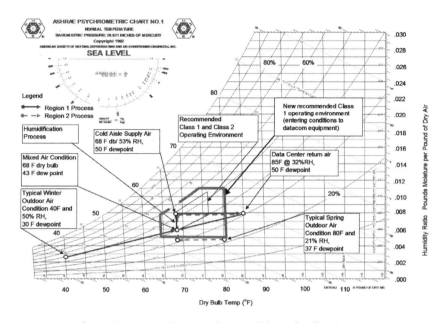

Figure 4.3a Sample economizer cycle conditions for Regions I and II (see Figure 4.3b for regions).

setpoint (in this case 68°F [20°C]). In the Figure 4.3a example, the mixed air will also need to be humidified from the 43°F (6°C) mixed air dew point up to the desired data center dew point of 50°F (10°C) before it is injected into the cold aisle. The return air temperature in this example is 85°F (29°C), also at a 50°F (10°C) dew point. (Return temperatures in hot aisles of datacom facilities are often significantly higher than the return temperatures in office environments due to the significant temperature rise of air passing through the IT equipment.)

Region II in Figure 4.3b represents the range of "spring/fall" ambient conditions, where air-side economizers introduce 100% outdoor air in order to minimize mechanical cooling requirements. In the "spring" example shown in Figure 4.3a, air at 80°F (26.7°C) and 21% RH is first sensibly cooled to 68°F (20°C) dry bulb, and then humidified to obtain the desired supply air relative humidity level.

Since the COP of the humidification process is different than the COP of the cooling process associated with cooling the return airstream, energy calculations must be performed to make sure that activation of the economizer cycle actually results in less energy consumption for a specific outdoor air temperature and humidity

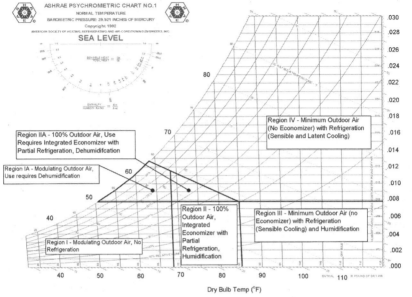

Figure 4.3b Air-side economizer environmental regions.

condition. Since humidification can also utilize natural gas or other nonelectric energy sources, fuel costs also need to be considered to minimize TCO.

Region IA of Figure 4.3b has a temperature lower than the cold-aisle supply temperature, but the dew point of this outdoor air is higher than the desired dew point of the supply air. The decision to use some percentage of outside air (i.e., the economizer) for operation in Region IA will depend on the energy cost associated with achieving the desired supply air condition with and without economizer operation. A hybrid desiccant dehumidification system may be the most efficient approach to achieving the required dehumidification, as there is no need to cool the air down to saturation for cooling purposes (and to cool the air to saturation purely for dehumidification purposes would have a low COP due to the unnecessary sensible cooling). While they do not specifically address data center dehumidification, several recent papers address strategies that utilize the hybrid desiccant dehumidification concept (Henderson et al. 2003; Yong et al. 2006; Saidi et al. 2007).

Operation in Region IIA will be similar to operation in Region IA. The exact temperature and humidity condition of the outdoor air needs to be determined, and the energy cost of conditioning that air must be compared to the cost of cooling return air to the same condition. Simply comparing the enthalpy of the ambient versus return air with a dual-enthalpy controller is not sufficient in this case because the COP of the conditioning process utilized for dehumidifying the ambient air will likely be different than the COP of the 100% sensible cooling process for conditioning the return air. The term *dual enthalpy* refers to the real-time measurement of both outdoor air and return air enthalpy. It is typically used to ensure that the enthalpy of the outdoor air is lower than the enthalpy of the return air before the economizer cycle is activated.

In Regions III and IV of Figure 4.3b, air-side economizers close outdoor air dampers to the minimum positions needed for data center pressurization or ventilation requirements. Introduction of minimum outdoor air in Region III would require sensible cooling and humidification, while Region IV ambient conditions would require both sensible and latent cooling.

There are areas within Region III where the enthalpy of the outdoor air is less than the enthalpy of the return air, but care should be taken that the enthalpy controller does not switch over to "low enthalpy" outdoor air at a dry bulb above the return air temperature. Since there is essentially no moisture gain within a data center, a latent cooling load on the coil from return air does not exist, and the use of outdoor air at a higher temperature than the return temperature will result in increased energy consumption.

The air-side economizer analysis shown in Figure 4.3 is somewhat simplified in that an actual data center typically has a recommended range of temperature and humidity (see Figure 2.2a), rather than a specific design point. The relative humidity dead band between Regions III and IV, where neither humidification nor dehumidification is required, would correspond to 41.9°F dew point–60% RH and 59°F dew point for ASHRAE Class 1 and Class 2 data centers. (It is important to note that the temperature and humidity conditions in Figure 2.2a are inlet conditions to the datacom equipment.) The return air temperature to the air handler may be considerably higher, since the typical temperature rise through datacom equipment is 20°F (11.1°C) or more. A dual enthalpy control system can adjust the regions shown in Figure 4.3 to the actual return air operating point, and help to optimize economizer cycle savings. (A single enthalpy controller only measures the return air enthalpy, and presumes a return air setpoint, but this is not typically a good assumption in a data center.) The full benefits of enthalpy economizers depend on accurate calibration and maintenance of the humidity sensors. These sensors require calibration more frequently than temperature sensors. Without adequate maintenance, enthalpy economizers have the potential to waste large amounts of energy due to incorrectly positioned dampers and can also negatively affect the environmental conditions in the facility.

Historically, there have been four concerns with the use of air-side economizers in datacom facilities. The four concerns are:

1. Increased particulate contamination and/or increased maintenance cost on filters.
2. Increased gaseous contamination.
3. Loss of humidity control during economizer operation and loss of a vapor seal during non-economizer operation.
4. Temporary loss of temperature control during economizer switchover with nonintegral economizers.

A recent study by LBNL has examined the first and third concerns (LBNL 2007b). It found that particulate concentrations in data centers without air-side economizers were about a factor of 10 below recommended maximum limits. Values increased during economizer operation, but if MERV 11 (ASHRAE Class II 85%) outdoor air filters are substituted for the MERV 7 filters (ASHRAE Class II 40%), it should be possible to reduce particulate concentrations during economizer operation to those of current data centers that do not use air-side economizers. The addition of filters with higher filtration efficiency will typically increase fan energy, and this will act to reduce the net energy savings of air-side economizers. With proper design and good maintenance, however, the increased pressure drop can be kept to a reasonable value. Applicable ASHRAE standards for air filtration include ASHRAE Standard 52.1-1992 and ASHRAE Standard 52.2-1999.

Relative humidity control with economizer operation was also measured and reported in the LBNL report. For one of the RH tests, the space relative humidity stayed within ASHRAE's recommended range (41.9°F dew point–60% RH and 59°F dew point), while for the other the humidity was on the low side of the recommended range but still mostly within the allowable range (20%–40% RH).

Several recommendations were made by the report for future work, especially for additional research on particulates and their effects on datacom equipment. When using an air-side economizer, the effect on gaseous contaminants in the data center, and specifically their effect on IT equipment, needs to be further studied. Some studies have shown that when corrosive gases have been exposed to IT equipment, catastrophic failure has occurred.

4.4 ADIABATIC-COOLED AIR-SIDE ECONOMIZERS

A derivative of the air economizer discussed above that is more energy efficient is one that uses an adiabatic wetted media humidifier/cooling component. This device is often referred to as a direct evaporative cooler. Figure 4.4 shows a central station air-handling unit schematic with the evaporative cooling component located upstream of the refrigeration cooling coil and a blow through supply fan arrangement. This configuration will optimize the humidification/cooling contribution of the adiabatic device. The plenum-type supply fan will typically ensure good blending of the outdoor air and return air volumes. Environmental considerations dictate proper treatment of the wetted media water supply (Lentz 1991) to minimize the impact of mildew, algae, and other issues that may cause health concerns.

Figures 4.5a and 4.5b illustrate regions of ambient conditions that will allow the adiabatic device to act as an automatic dew-point control for the air supplied to the data center. By modulation of outdoor air and return air dampers, the heat generated

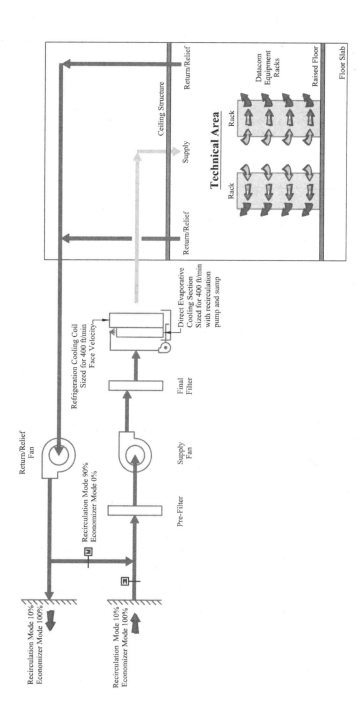

Figure 4.4 Air-handling system with air-side economizer and evaporative cooling (see Figures 4.5a and 4.5b for process).

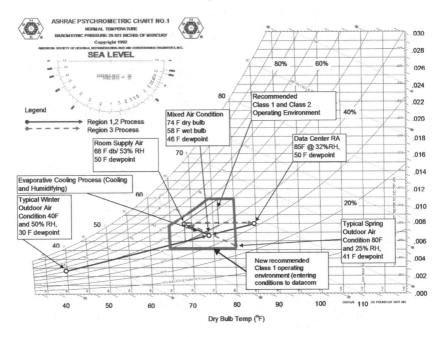

Figure 4.5a Sample evaporative economizer conditions for Regions I and III (see Figure 4.5b for regions).

inside the data center is blended with outdoor air to a wet-bulb condition which, when introduced to the direct evaporative cooler, will provide the proper supply air temperature and humidity condition.

A wetted rigid-media direct evaporative cooling device should be selected at approximately 400 ft/min (2.03 m/s) face velocity. The saturation efficiency at that face velocity for a 12 in. (0.3 m) deep wetted rigid media is usually 90%, and the static pressure loss is in the range of 0.15 in. w.g. (37.4 Pa) at sea level (ASHRAE 2004d). An alternative method of evaporative cooling that will perform the same adiabatic cooling and humidification function is the atomizing-type humidifier.

Region I in Figure 4.5b encompasses the ambient temperatures below 64.4°F (18°C) dry bulb when both cooling and humidification of the data center is required. Both may be achieved without energy expenditure for either refrigeration or humidification.

Region II represents outdoor conditions where cooling without refrigeration may be extended past the 64.4°F (18°C) dry-bulb threshold. Note that in Figure 4.3b for the air economizer, Region II, without the direct evaporative cooling device,

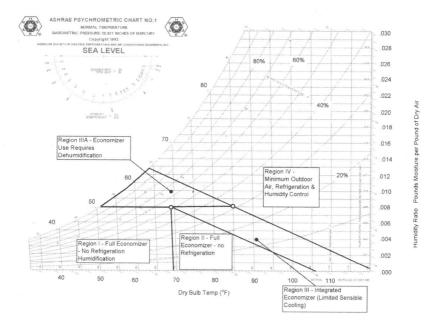

Figure 4.5b Direct evaporative economizer air-side economizer environmental regions.

requires both mechanical cooling and humidification to deliver the 68°F (20°C) dry bulb and 50°F (10°C) dew point to the space.

Region III shows ambient conditions above 58°F (14°C) WB and less than 64°F (18°C) wet bulb where the direct evaporative cooling device will help minimize mechanical cooling requirements.

Economizer operation in Region IIIA requires more careful analysis. The enthalpy of the outdoor air in this region is less than the enthalpy of the return air, but dehumidification is required if the economizer cycle is enabled. The energy cost of dehumidification of the outdoor airstream may exceed the energy required for cooling of the return airstream, since the COP of the economizer conditioning process—which will likely include desiccant dehumidification—will likely be lower than the COP of the return airstream cooling process—a sensible cooling process that can be achieved with mechanical refrigeration.

Region IV represents ambient dry-bulb and wet-bulb conditions when the outdoor air dampers should be set at their minimum (i.e., economizer air shutoff) to minimize refrigeration energy.

4.5   WATER-SIDE ECONOMIZERS

Water-side economizers use cool outdoor dry-bulb or wet-bulb conditions to generate condenser water that can partially or fully meet the facility's cooling requirements. There are two basic types of water-side economizers: direct and indirect free cooling. In a direct system, the condenser water is circulated directly through the chilled-water circuit. In an indirect system, a heat exchanger is added to separate the condenser-water and chilled-water loops. For comparison, Figure 4.6 shows a schematic of a direct water-side economizer, and Figure 4.7 shows a schematic of an indirect water-side economizer. The direct water-side economizer is the most efficient, and maximizes the number of annual economizer hours, but operations can be more complex, and there may be tube-fouling concerns associated with running condenser water through cooling coils.

Water-cooled chilled-water plants incorporating water-side economizers often use the indirect approach. In this approach, the chilled-water plant cooling towers generate cold condenser water, which is then passed through a heat exchanger where it absorbs the heat from the chilled-water loop. When the condenser water is cold enough such that it can fully meet the cooling load, the water chillers can be shut off.

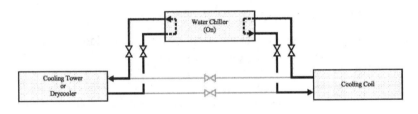

NORMAL CHILLER PLANT OPERATION (CONDENSER WATER TEMPERATURE ABOVE COIL DESIGN TEMPERATURE)

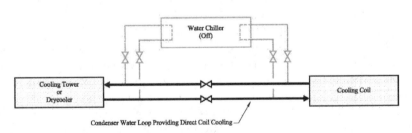

DIRECT WATER-SIDE ECONOMIZER OPERATION (CONDENSER WATER TEMPERATURE BELOW COIL DESIGN TEMPERATURE)

Figure 4.6   Direct water-side economizer.

In order to fully meet data center cooling loads, outdoor wet bulbs generally have to be 7°F–10°F (4°C–6°C) less than the design chilled-water temperature, depending on the approach temperature of the tower and the design temperature rise across the heat exchanger.

Water-side economizers used in conjunction with central chilled-water plants can be either arranged in series or parallel. When arranged in parallel, the chillers are shut off and the economizers' heat exchanger enabled when the outdoor wet-bulb temperatures are low enough to allow full cooling through the heat exchanger. When arranged in series (as shown in Figure 4.7), chilled water is routed through the heat exchanger whenever the ambient wet-bulb temperature is low enough to allow a portion of the cooling load to be met by the heat exchanger, thus complying with the integrated economizer requirement—Section 6.5.1.3 of ASHRAE Standard 90.1 (ASHRAE 2004a). In this mode, chilled water flows through both the heat exchanger and chiller in series, resulting in a significant number of partial "free cooling" hours, with the only additional energy expense being the energy required to pump water through both the heat exchanger and the chiller. Once the wet-bulb

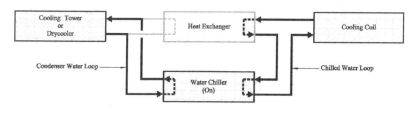

NORMAL CHILLER PLANT OPERATION (CONDENSER WATER TEMPERATURE ABOVE COIL DESIGN TEMPERATURE)

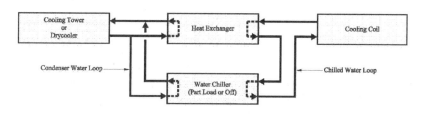

INDIRECT WATER-SIDE ECONOMIZER OPERATION (CONDENSER WATER TEMPERATURE BELOW COIL DESIGN TEMPERATURE)
(INTERGRATED ARRANGEMENT SHOWN IN THIS DIAGRAM, ALLOWING FOR PARTIAL ECONOMIZER OPERATION AS A PRE-COOLER)

Figure 4.7 Indirect water-side economizer.

conditions permit full cooling by the heat exchanger, flow can bypass the chiller evaporator, reducing pumping head and increasing energy efficiency.

Water-side economizers, when used in conjunction with water-cooled chilled-water plants, require heat exchangers, larger cooling towers, and possibly additional pumps (dependent upon the piping arrangement). Additionally, the economizers require motorized control valves to divert the direction of the water flow to the heat exchanger as required. These changes of valve position, if programmed to perform automatically, are likely to occur often in a moderate climate. Frequent changes of state may need to be monitored to ensure reliable transition from compressor cooling to free cooling.

Direct water-side economizer operation can take a number of forms. One of the most common in the data center environment integrates independent coils within the CRAC unit to allow cooling from either of two sources One is for free cooling and other is the DX coil, as shown in Figure 4.8. The free cooling coil is connected to a cooling tower, a dry cooler, or a spray-cooled dry cooler. The economizer operates when outdoor dry-bulb conditions can cool the condenser water to a temperature that can support partial or full cooling of the data center. The cool condenser water is then passed directly through an auxiliary cooling coil in the CRAC that absorbs heat from the facility. A schematic of this strategy has already been introduced in Figure 3.12. A "wet cooler," which is similar to a dry cooler but with a water spray directed over the heat transfer surface, can increase the duration of the free cooling by lowering the temperature of the condenser water during many hours of the year. One consideration when using this approach, if the cooling tower is an open loop tower, is to utilize special CuNi in the coils to minimize the impact of contaminants within the cooling fluid that can cause erosion and pitting within standard copper coils, resulting in premature leaks within the coil.

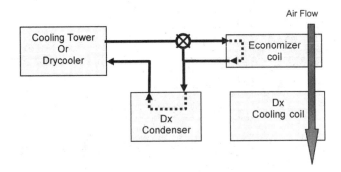

Figure 4.8 Water-side economizer when integrated into a CRAC unit.

4.6   CLIMATIC ADVANTAGES OF VARIOUS ECONOMIZERS

Mission critical applications with 24/7 duty cycles offer a significant potential for mechanical cooling energy reductions depending on climatic conditions at the site. Table 4.1 uses binned joint frequency weather data for seven different locations to illustrate the percentage of annual hours within the six climatic regions depicted in Figure 4.3b. Binned joint frequency data is a matrix that provides information on the number of hours per year that a site is at a certain temperature and humidity—these data are available from many sources, including ASHRAE's *Weather Data Viewer* (ASHRAE 2005f.) The climates of Denver, Colorado, and Seattle, Washington, have the highest percentage of hours in Region I, where refrigeration energy reduction potential is greatest. When the hours in Regions I and II are combined, a dry climate such as Denver, Colorado, shows the highest potential for total annual refrigeration energy reduction, but it also suffers the largest humidification energy costs when large amounts of outdoor air are introduced. Climates like those in Atlanta, Georgia, and much of the southeastern United States have a high percentage of hours in Region IV, above both 63°F wet bulb and 55°F dew point, when outdoor air should be reduced to a minimum and only dehumidification and mechanical cooling are required.

Table 4.2 looks at binned weather data in seven representative cities, and sorts the annual hours in each of the five regions of the wet-bulb economizer psychometric chart in Figure 4.5b. These data show that arid and cold climates have the biggest potential for refrigeration and humidification energy savings. The southeast, where ambient humidity is higher, would not realize the refrigeration ton-hour avoidance

Table 4.1   Air-Side Economizers—Utilization Potential for Selected Cities for Supply and Return Conditions Shown in Figure 4.3a*

	Los Angeles	San Jose	Denver	Chicago	Boston	Atlanta	Seattle
Region I	30.9	51.6	75.3	64.0	63.8	42.6	71.1
Region IA	47.6	26.4	5.0	12.3	15.5	15.1	18.5
Region II	3.0	5.4	13.1	2.5	2.2	3.6	3.2
Region IIA	8.0	10.4	2.5	4.6	4.4	4.6	5.7
Region III	0.2	0.9	3.7	0.1	0.1	0.1	0.2
Region IV	10.2	5.4	0.4	16.5	13.9	34.0	1.4

* Note: The above figures utilize 24/7 bin data, as is typical of datacom facilities.

Table 4.2 Adiabatic-Cooled Air-Side Economizer— Utilization Potential for Selected Cities for Supply and Return Conditions Shown in Figure 4.5a[*]

	Los Angeles	San Jose	Denver	Chicago	Boston	Atlanta	Seattle
Region I	30.9	51.6	75.3	64.0	63.8	42.6	71.1
Region II	2.7	3.4	11.6	1.6	1.5	2.5	1.8
Region III	0.5	2.6	5.2	1.1	0.8	1.2	1.5
Region IIIA	55.6	36.8	7.5	16.9	20.0	19.7	24.1
Region IV	10.3	5.6	0.4	16.5	14.0	34.0	1.4

* Note: The above figures utilize 24/7 bin data, as is typical of datacom facilities.

of colder and dryer climates. The Rocky Mountain western regions at high altitude would benefit the most from an adiabatic economizer.

When evaluating economizers, the reduction in cooling costs should always be compared with the potential added costs associated with maintenance, building space, and equipment reliability problems—true TCO evaluation. Water-side economizers should be considered in conjunction with air-side economizers. In making an evaluation and comparison, the following points should be taken into consideration:

- There is wide geographic (climatic) variability in the percentage of the year that economizers can be used.

- Some climates are better for water-side economizers and some are better for air-side economizers, depending on ambient humidity levels.

The percentage utilization of economizers (both water and air side) increases substantially if the data center can operate with a higher supply air temperature.

Table 4.3 lists the percentage of the year that adiabatic air and water-side economizers could potentially provide full cooling as a function of chilled-water supply temperature for selected cities in the United States. The data show that (1) there is wide geographic (climatic) variability in the percentage of the year that economizers can be used, (2) some climates are better for water-side economizers and some are better for air-side economizers, and (3) the percentage utilization of economizers increases substantially if the data center can operate with a higher supply air temperature.

Table 4.3 Relative Availability of Water-Side and Air-Side Economizer Hours for Selected US Cities as a Function of Supply Air Temperature

Water-Side Economizer Hours with No Required Mechanical Cooling*,†

Outdoor Air Wet-Bulb Bin, °F (°C)	CWS,* °F (°C)	Supply Air Temperature,† °F (°C)	Los Angeles % of Year below Wet Bulb	San Jose % of Year below Wet Bulb	Denver % of Year below Wet Bulb	Chicago % of Year below Wet Bulb	Boston % of Year below Wet Bulb	Atlanta % of Year below Wet Bulb	Seattle % of Year below Wet Bulb
59 (15)	66 (19)	70 (21)	68%	78%	93%	75%	75%	56%	90%
53 (12)	60 (16)	64 (18)	36%	46%	77%	64%	63%	44%	68%
47 (8)	54 (12)	58 (14)	13%	21%	63%	55%	52%	33%	45%
41 (5)	48 (9)	52 (11)	3%	6%	51%	46%	41%	22%	21%

Air-Side Economizer Hours with No Required Mechanical Cooling‡,**,††

Outdoor Air Dry-Bulb Bin, °F (°C)	Supply Air Temperature,‡ °F (°C)	Los Angeles % of Year below Dry Bulb	San Jose % of Year below Dry Bulb	Denver % of Year below Dry Bulb	Chicago % of Year below Dry Bulb	Boston % of Year below Dry Bulb	Atlanta % of Year below Dry Bulb	Seattle % of Year below Dry Bulb
69 (21)	70 (21)	86%	80%	82%	80%	83%	65%	65%
63 (17)	64 (18)	59%	64%	72%	70%	71%	51%	51%
57 (14)	58 (14)	32%	39%	61%	62%	61%	41%	41%
51 (11)	52 (11)	6%	18%	51%	52%	50%	29%	29%

Table 4.3 Relative Availability of Water-Side and Air-Side Economizer Hours for Selected US Cities as a Function of Supply Air Temperature (continued)

Adiabatically Humidified/Cooled Air-Side Economizer Hours with No Required Mechanical Cooling[‡,**,††]

Outdoor Air Wet-Bulb Bin, °F (°C)	Supply Air Temperature,[†] °F (°C)	Los Angeles % of Year below Wet Bulb	San Jose % of Year below Wet Bulb	Denver % of Year below Wet Bulb	Chicago % of Year below Wet Bulb	Boston % of Year below Wet Bulb	Atlanta % of Year below Wet Bulb	Seattle % of Year below Wet Bulb
69 (21)	70 (21)	99%	100%	100%	93%	95%	82%	100%
63 (17)	64 (18)	87%	93%	99%	83%	85%	65%	98%
57 (14)	58 (14)	53%	70%	89%	71%	72%	51%	85%
51 (11)	52 (11)	23%	39%	73%	60%	60%	39%	62%

* Chilled-water supply temperature assumes a 7°F (4°C) approach between wet-bulb temperature and chilled-water supply temperature for all conditions. Cooling tower selections need to be checked for an actual application.

† Supply air temperature assumes a 4°F (2.2°C) approach between condenser-water temperature and supply air temperature for all conditions. A heat exchanger needs to be engineered for an actual application.

‡ Supply air temperature assumes a 1°F (0.6°C) temperature rise to account for fan heat and friction heat rise.

** The percentage listed in this table refers to the ability of outside air to meet the listed temperature specification. Humidification and dehumidification will also be needed to meet recommended ASHRAE conditions.

†† Caution: Owners have been known to disable air-side economizers in data centers.

4.7   ENERGY-EFFICIENCY RECOMMENDATIONS/BEST PRACTICES

Economizer cycles provide an opportunity for substantial energy savings and increased TCO in datacom facilities. The primary energy savings associated with economizer cycles is reduced mechanical cooling. The following are the most important considerations in choosing an economizer:

*   Climate is the most important variable in comparing air-side and water-side economizer savings. Table 4.3 shows that the savings are comparable for both approaches, but for some sites air-side economizers are best, while for others water-side economizers will yield the highest savings.
*   Raising the supply air setpoint in the datacom facility can have a dramatic effect on the number of partial and full economizer hours available in a facility and, thus, on annual energy consumption of the chiller plant.
*   Integrated economizers, which allow for partial use of the economizer cycle (whether air or water), should be used to maximize economizer savings.
*   Adiabatic air-side economizers can be more energy efficient than "standard" air-side economizers, but utilization of this technology is low for datacom facilities at this time.
*   For air-side economizers, the COP of the dehumidification process, and the type and fuel source of the humidification process, are important parameters. Humidification and dehumidification is much less of an issue with water-side economizers.

5

Airflow Distribution

5.1   INTRODUCTION

It is fair to say that adequate airflow distribution in datacom facilities (particularly those with high-density equipment) is one of the biggest challenges of datacom facility design. Airflow design clearly impacts the "HVAC Fans" slice of the energy pie shown in Figure 5.1, but it also can substantially impact the "HVAC Cooling" energy, and, thus, both slices of the pie are highlighted, as will be apparent after reading this chapter.

Average Data Center Power Allocation

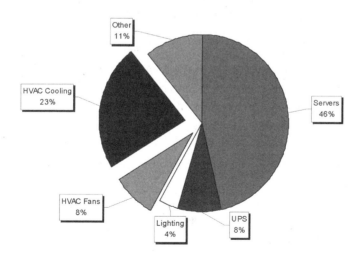

Figure 5.1   Energy consumption impact of airflow distribution (LBNL average of 12 data centers.

Where one draws the line to define a point of sufficient cooling for a space has a huge impact on the energy consumption and TCO. But where should this line be drawn? For occupied spaces, one only has to be able to define human comfort, and cool to that condition. A good source for defining the envelope of human comfort is *ANSI/ASHRAE Standard 55-2004, Thermal Environmental Conditions for Human Occupancy* (ASHRAE 2004j). For critical facilities, such as data centers, how does one know when equipment is "comfortable"? This question is answered below by considering applicable environmental and equipment specifications, and then by assessing compliance with those specifications. How this equipment comfort envelope is defined and how air is distributed within a data center to cool the equipment has a huge impact on the energy consumption of a facility. The remainder of this chapter addresses the airflow issues, which clearly have a large impact on efficient energy usage and TCO.

1. *Environmental Specifications.* In the United States, ASHRAE and Telcordia have both developed environmental specifications addressing equipment comfort in data spaces. Telcordia's (formerly Bellcore) NEBS GR-3028-CORE provides environmental specifications most appropriate for telecommunications facilities such as central offices; *Thermal Guidelines for Data Processing Environments* (ASHRAE 2009) is targeted more toward establishing environmental specifications for data centers. A strong case can be made that compliance to an accepted specification (ASHRAE's and Telcordia's are not the only two—there are others) correlates to low risk of equipment failure and, thus, down time. While the huge energy usage of today's data centers creates strong incentive to reduce energy usage, conformance with environmental specifications is the first order of business in any critical facility's design. Energy usage and/or TCO can only be considered after the conformance with the environmental specifications is guaranteed.

2. *Measurement of Level of Conformance.* To conform precisely to *Thermal Guidelines for Data Processing Environments* (ASHRAE 2009), the inlet temperature and relative humidity for *all* equipment within a data center must fall within an appropriate range. Such a black or white test for conformance gives no indication of how well a data center conforms (or does not conform). Herrlin (2005a) proposed the rack cooling index (RCI), which can be used in at least two ways. When used to describe an existing facility, it provides a measure of how closely actual environmental conditions approach the

prescribed specifications. For a new facility, RCI can be used to specify the level of conformance required as part of a Basis of Design document.

5.2 AIRFLOW MANAGEMENT AT THE DATA CENTER LEVEL

The choice of cooling system architecture directly affects energy consumption through fan power (the energy required to deliver cooling air to and capture warm exhaust air from equipment) and cooling power (the power required to cool the warm rack exhaust air back to the required supply temperature). With fan power, the greater the pressure drop the fan must overcome, the greater the energy consumption. With cooling power, within some practical limits, the greater the degree of separation between the cool supply and the warm return streams, the greater the operating efficiency of the cooling units.

Cooling system architecture affects TCO beyond the initial cost of the system. For example, computing throughput is often limited by the amount of cooling that can be delivered to a particular location of the data center. It is also possible for the cooling system to be capable of cooling equipment densities that will never be realized. Cooling system architecture also affects TCO through floor-space considerations, serviceability and manageability issues, etc. A best practice is to right-size cooling systems to match the actual load with planned and installed flexibility to incrementally add capacity as loads increase over the life of the facility.

As discussed in *Design Considerations for Datacom Equipment Centers* (ASHRAE 2005d), the dimensionless indices return heat index (RHI) and supply heat index (SHI) proposed by Sharma et al. (2002) quantify the degree to which the cold and hot streams are separated from one another. These indices, along with rack inlet temperatures, are often computed through CFD analysis.

5.2.1 Vertical Underfloor (VUF)

Figure 5.2 shows a data center with a vertical underfloor (VUF) cooling architecture. Cooling air is supplied throughout the data center via a raised-floor plenum and ultimately to equipment racks through perforated floor tiles. Warm exhaust air is typically returned "through the room" to cooling units that are often located around the perimeter of the room. Warm air is typically drawn into the tops of the CRAC or CRAH cooling units; cool air is supplied into the floor plenum.

Reliable and uniform air distribution may be difficult to achieve because of a nonuniform pressure distribution in the plenum created by insufficient plenum depth, the presence of blockages, irregular layouts, insufficient perforated-tile resistance (i.e., highly open tiles), and excessive leakage through cable cutouts and other

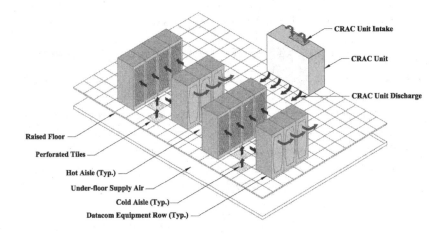

CRAC Unit Intake

CRAC Unit

CRAC Unit Discharge

Raised Floor

Perforated Tiles

Hot Aisle (Typ.)

Under-floor Supply Air

Cold Aisle (Typ.)

Datacom Equipment Row (Typ.)

Figure 5.2 Example data center with VUF cooling architecture.

openings. (See VanGilder and Schmidt [2005] for more information.) However, the pressure drop through the cooling unit is generally many times greater than the static pressure in the plenum. Consequently, cooling unit fans are not significantly affected by plenum pressure, and there is generally little opportunity for saving fan power by decreasing the resistance in the plenum.

With respect to cooling power, higher efficiency can be achieved by providing better separation of cool and warm streams. To achieve a reasonable degree of separation, a hot/cold aisle protocol should be adopted (ASHRAE 2009), blanking panels should be installed in all open rack locations, airflow leakage through the raised floor should be minimized, and steps should be taken to ensure fairly uniform airflow delivery through all supply air openings. Furthermore, careful attention should be paid to the airflow return paths and resulting temperatures to ensure that all cooling units carry a reasonable cooling load; a useful strategy is to extend the returns of the cooling unit with ductwork so that all cooling units draw the warmest air from close to the top of the space.

Other TCO considerations related to a VUF architecture include the loss of data center floor space to cooling units and ramps, installation and maintenance costs associated with the raised floor and ceiling plenum (where utilized), and engineering costs associated with the room's unique geometry and mission.

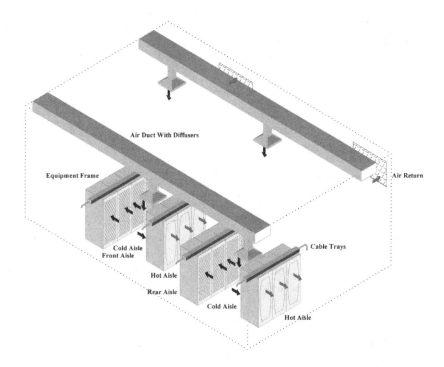

Figure 5.3   Example data center with VOH cooling architecture.

5.2.2   Vertical Overhead (VOH)

Figure 5.3 shows a data center with a VOH cooling architecture. Cooling airflow is typically supplied to the equipment racks through registers via a manifolded system of ductwork connected to air handlers. Airflow returns are typically mounted near the top of the space. Where the air handlers are located directly on the data center floor, a partial height wall can be built around the air handlers to encourage the warmer air from the top of the space to enter into the return air openings of the air handlers. With VOH systems, cool supply air typically mixes with the warmer surrounding room air before reaching the equipment racks. Consequently, rack inlet temperatures are generally higher than supply temperatures, although they are often more uniform relative to a corresponding VUF system. By contrast, the VUF system typically provides equipment mounted lower in the racks with airflow near the supply temperature; however, equipment mounted higher in the racks may receive substantially warmer air, including exhaust air recirculated over or around the racks.

Because of the ductwork, the static pressure on the fans of the air handlers is typically higher than that seen by the cooling units in a VUF system. However, because of the greater uniformity of inlet temperatures, it may be possible to achieve better performance with less airflow when compared with a VUF system.

As with a VUF system, higher cooling efficiency can be achieved by providing better separation of cold and hot streams.

Other TCO considerations related to a VOH system include the installation costs of the ductwork, planned future changes to the room layout, and engineering costs associated with the room's unique geometry and mission.

5.2.3   Local Cooling

Recently, cooling systems have been developed in which the cooling units are located close to the equipment racks, either within the rows of racks themselves or mounted over the tops of the racks. Such local cooling provides closer coupling between equipment needs and cooling resources, and is inherently scalable. Because of the proximity to equipment racks, fan power is fairly low, and there is typically good separation of hot and cold streams.

One example of local cooling is a self-contained rack/cooling system. Such a system allows for precise control of server inlet temperatures; however, the method of operation during a power loss disturbance or other non-normal cooling condition needs to be analyzed.

5.2.4   Supplementary Cooling

Supplementary cooling devices may be located within or on a rack, and include products that increase the cool air supply to a rack, capture the hot air exhaust from a rack, or partially cool the air leaving the rack. Fan-only devices may be an effective way to manage local hot spots and help improve separation between the cool and warm streams; however, such units do not provide additional net cooling capacity. In fact, the power required to run the fans adds a small amount to the overall room cooling requirements. Consider an example in which all racks in a facility are fitted with supplemental air-moving products that capture all of the rack exhaust air and duct it back to the cooling units via a return plenum. In this case, substantial fan power load has been added to the room; however, the efficient separation of cool and warm streams may yield greater efficiency—possibly to the point where fewer cooling units (and associated cooling unit fan power) are required.

Unlike the fan-only products, the true supplemental cooling units add to the net cooling capacity of the room and, therefore, may reduce the number or capacity of traditional cooling units or air handlers. For more details and an example TCO analysis, see Schmidt et al. (2005).

5.3   AIR HANDLER/CRAH/CRAC UNIT SELECTION

5.3.1   Operating Pressure

The electrical energy consumption of a fan motor, all other factors being equal, is directly proportional to the pressure drop across the fan, so savings can be achieved by minimizing the pressure drop of an air-handling system. The operating pressure of fans is a function of both the internal and external components of the air handler or CRAH/CRAC unit, so both internal and external pressure drops should be analyzed and optimized. An example of an internal component pressure drop (in this case, a cooling coil) can be found in Figure 3.3 of Chapter 3. The external pressure drop would be the sum of the supply plenum and/or the supply ductwork, and the return plenum and/or return ductwork. For the purposes of this chapter, it is sufficient to say that the external pressure drop found in plenums is typically less than the pressure drop found in ductwork due to the lower air velocities and associated friction losses. All other factors being equal, a plenum supply and/or return will therefore have lower energy consumption than a ducted system.

5.3.2   Fan Selection

From the standpoint of energy efficiency, the choice of fans can be an important parameter. There are several types of fans (such as forward curved, backward inclined, and airfoil), and selection of the most efficient fan can take time. In general, higher airflow rates allow for the selection of fans with higher efficiency. Motor and VFD components also tend to be more efficient with larger fan systems. More information on fans can be obtained from Chapter 18, "Fans," in the *2004 ASHRAE Handbook—HVAC Systems and Equipment Volume* (ASHRAE 2004c).

5.3.3   Variable-Speed Fans

The energy-saving potential of variable-speed fans is well documented in the literature, and is also discussed in Chapter 3. Mention is made of the use of variable-speed fans in this chapter mostly as a warning. The balance between supply airflow and datacom equipment intake can be a delicate one, and changing that balance through the use of a variable-speed system should be closely monitored to make sure that datacom equipment inlet air conditions are maintained throughout the range of airflows of the variable-air-volume system. By the same token, most types of datacom equipment contain variable-speed fans for internal cooling purposes, so variation in overall system flow rates will occur even if the supply air equipment is maintained at a constant volume.

A good way to control the fan speed for optimal energy performance is to control the VFD output to keep a constant underfloor plenum (or duct) pressure. The addition or removal of tiles in a VUF configuration is typically a manual process coupled with the addition or removal of cabinets and/or loads. Despite the need to manually manage these tiles, the energy savings associated with allowing the fans to vary according to the actual loads provides a significant operating cost benefit to the owner.

5.4   COIL SELECTION/SUPPLY AIR TEMPERATURE

5.4.1   Selection of Supply Air Temperature

The selection of a supply air temperature must be done to comply with the design criteria of the datacom equipment. The inlet air temperature specification is typically in the range of 64.4°F–80.6°F (18°C–27°C) per *Thermal Guidelines for Data Processing Environments* (ASHRAE 2009).

Typical HVAC supply air temperatures, which almost always mix with room air, are usually in the range of 55°F–60°F (13°C–16°C), with the coil discharge temperature usually determined by the coil's dual function to serve as both a cooling and a dehumidification device. In a data processing environment, however, there is a *theoretical* possibility (if there is no mixing of supply and return airstreams) to have 64.4°F–80.6°F (18°C–27°C) air from a supply air system delivered to the inlet of the datacom equipment, and to satisfy the inlet conditions specified in *Thermal Guidelines for Data Processing Environments* (ASHRAE 2009). In practice, ideal delivery without mixing is difficult, and a more typical supply air temperature off of a cooling coil is in the range of 60°F–70°F (16°C–21°C). A higher supply air temperature typically results in (1) more efficient chiller operation, and (2) an increased number of economizer hours.

It may be possible to have lower supply airflow (and, thus, reduced fan energy) if a lower supply air temperature is chosen and the supply air is designed to mix with warmer room air prior to delivery to the datacom equipment. This is more typical of overhead delivery. The energy efficiency increase associated with reduced airflow, however, must be balanced against less efficient chiller operation and a decrease in economizer hours.

5.4.2   Chiller Efficiency/Pumping Power

Chiller efficiency is typically higher if the discharge temperature of the chilled water is increased. This has been illustrated in Figures 3.5a and 3.5b. Chilled-water pumping power is typically proportional to the quantity of chilled water pumped in the system, so choosing a cooling coil for a higher ΔT will decrease pumping system energy consumption. The water pressure drop across the cooling coil will also affect pumping power, so coil selections with a large pressure drop should be avoided unless they result in an air-side pressure drop with lower overall energy consumption. Figure 3.3 shows the relationship between air-side pressure drop and approach temperature for a typical coil.

5.4.3   Use of Economizers

Coil selection can have a significant impact on the percentage of the year that an economizer can operate. The two most important parameters are (1) supply air temperature, and, (2) for the case of water-side economizers, the approach temperature between the chilled-water supply temperature and AHU or CRAC supply air temperature. In the case of supply air temperature, a higher supply temperature will allow for a greater number of economizer hours in all climates. The exact percentage increase must be calculated based on an analysis of weather data.

5.4.4   Humidity Control

Relative humidity can be difficult to control to recommended tolerances in a datacom environment, particularly if there is a wide variation in supply air temperatures. In general, a wider operating range for relative humidity will decrease energy consumption: less humidification will be required in the winter, and less dehumidification energy will be required in the summer. The wider range may also help to avoid "fighting" between adjacent air handlers. In some cases, an adiabatic humidifier may be appropriate, which will both humidify and cool the supply air (see Section 4.4). It may also make sense to have an independent dehumidification system in order to allow the primary cooling coils to run dry. This will typically allow for more flexibility in the choice of a coil supply air temperature, which, in turn, may help to optimize airflow distribution. Humidity control issues are covered in more detail in Chapter 6 of this book.

5.5   AIRFLOW MANAGEMENT ADJACENT TO AND WITHIN THE DATACOM EQUIPMENT RACK

With respect to cooling capability, higher efficiency can be achieved by providing better separation of cool and warm airstreams. To achieve a reasonable degree of separation within a hot/cold aisle protocol (ASHRAE 2009), air management adjacent to and within the datacom equipment rack must be considered. If the airflow paths of the space and the datacom equipment are aligned, such as in a front-to-rear arrangement, a high degree of separation is typically obtained. In cases where airflow directivity of the space and the datacom equipment are not aligned, however, specialized analysis is typically called for to ensure high efficiency.

5.5.1   Rack Airflow Configurations

Rack airflow configurations, such as front-to-rear and front-to-top, can play an important role in datacom energy efficiency. The front-to-rear configuration is the most commonly employed in rack enclosures since it most easily facilitates the hot-aisle/cold-aisle layout.

Other configurations, such as side-to-side and bottom-to-top, can be employed to accommodate the specific requirements of the datacom equipment. In addition, some rack enclosures employ fan-assisted intakes, with and without fan-assisted exhausts. Each of these various configurations may improve the separation between the hot and cold airstreams. However, consideration must be given to the trade-off between improving the airflow effectiveness and increasing the fan energy associated with the air movement through the racks. The design engineer and the IT professional should plan the space layout and selection of the rack enclosures together in order to ensure an energy-optimized configuration.

5.5.2 Blanking Panels

Higher efficiencies are achieved when a higher degree of separation between the cold air supply and hot air return can be maintained. Blanking panels placed in open locations in a rack will prevent the datacom equipment's hot exhaust air from the back to the front inlet of the datacom equipment. When open space is not blocked by the use of blanking panels, datacom equipment fans uniformly pull air from the front of the rack, as well as pull heated exhaust air from the rear of the rack, adversely increasing the overall temperature at the inlet of the datacom equipment.

5.5.3 Raised Floor Cable Cut-Out Sealing Grommet

With a VUF system, conditioned air is delivered to the inlet of the datacom equipment through the raised floor plenum. A cut-out in the raised floor for power or data cables allows air generated by the cooling unit(s) to bypass the inlet of the datacom equipment. The amount of conditioned air generated by the cooling unit that bypasses the datacom equipment through cable cut-outs can be significant and a significant improvement in efficiency is achieved when the cable cut-out is blocked by the use of a sealing grommet or pillow. Careful sealing of other raised floor openings and holes in perimeter walls is also important to maintain good cooling efficiency.

5.6 AIRFLOW OPTIMIZATION

Air is the main carrier of heat and moisture in the data center. It is important to optimize the flow paths of both cold supply air and hot return air in the data center. The goal of such optimization should be to minimize unnecessary mixing of these two streams, including reducing short-circuiting of cold air back to the air-conditioning units and short-circuiting of rack exhaust air back to rack inlets. Both of these processes can adversely affect the energy efficiency of data centers. Care must also be taken (with a VUF arrangement) to consider and/or avoid negative air pressures underfloor near CRAH/CRAC discharges. Consideration should be given for such airflow optimization both at the initial design stage as well as on regular basis during the operational phase. The following are some of the optimization schemes that can be employed:

1. *Best Practices Using Guidelines from Benchmarked Data.* Initial design of data center layout and equipment selection should be based on published guidelines for best practices. Please refer to *Design Considerations for Datacom Equipment Centers* (ASHRAE 2005d) A few important guidelines are listed below.

    *   For raised floor data centers, separation of cold and hot aisles help in keeping the two airstreams separated without mixing.
    *   In addition, provision of an overhead ceiling plenum return can provide a separate isolated path for the hot return air.
    *   Alignment of air-conditioning units at the end of the hot aisle also provide an easy path for hot return air back to the air conditioner and can avoid direct short-circuiting of cold air from cold aisles.

2. *CFD Modeling.* Computational fluid dynamics (CFD) analysis can be an effective tool for the optimization and troubleshooting of airflow. As mentioned before, an energy-efficient layout of a data center depends on several factors, including the arrangement and location of various data center components including the positioning of air-conditioning systems and positioning of racks and other components. CFD analysis, which is based on sound laws of physics, can help visualize airflow patterns and resulting temperature distribution in a data center facility. Any mixing and short-circuiting airflow patterns can be visualized through the airflow animations based on CFD analysis instead of intuitive imagination.

    Optimization of data center layout and equipment selection through CFD analysis at the design evaluation phase can help in avoiding future problems related to poor airflow distribution and resulting hot spots in the data center facility. CFD analysis can provide a detailed cooling audit report of the facility, including the performance evaluation of individual components, including air-conditioning units, racks, perforated floor tiles, and any other supplementary cooling systems in the facility. Such reports can help in the optimization, selection, and placement of various components in the facility. For example, such CFD-analysis-based cooling audit reports of air-conditioning systems can predict relative performance of each unit, indicating over- or underperformance with respect to other units on the floor, which can affect the energy efficiency and total cost of ownership long term. Performance of each component can be easily tested against the guidelines in *Thermal Guidelines for Data Processing Environments* (ASHRAE 2009) to meet the optimum requirements.

    CFD analysis can also help in predicting the nondimensional indices (RCI, RHI, and SHI) described earlier in this chapter. Testing and evaluation of these indices would require extensive measurements of several entities such as inlet temperatures at each server, and airflow measurements through each perforated floor tile, each server, and through each air-conditioning unit. CFD analysis, if performed at the design evaluation phase, can predict the overall values of these indices not only at the entire facility level but also at the individual component level.

5.7   ENERGY-EFFICIENCY RECOMMENDATIONS/BEST PRACTICES

Good airflow management is critical to an energy-efficient data center. The best practices recommended from this chapter are:

- Consider several possible airflow cooling architectures. Some may be better suited to a particular datacom equipment mix and/or layout than others. If there is any question about proper performance, a CFD study is recommended for analysis and optimization.
- Provide good separation between the supply and return air pathways. A hot-aisle/cold-aisle design is an effective approach for this separation. If downflow CRAH/CRAC units are used, they should be located at the end of the hot aisle.
- Select as high a supply air temperature as will provide sufficient cooling to the datacom equipment to maximize cooling equipment efficiency and the number of available economizer hours.
- Select energy-efficient fans, fan motors, and VFDs to keep fan energy consumption to a minimum.
- Recognize that datacom equipment loads will change over the next 10–15 years. Develop a cooling distribution strategy that can accommodate these changes.

6

HVAC Controls and Energy Management

6.1   INTRODUCTION

A properly designed, commissioned, and maintained control system will have a significant impact on energy costs and TCO in almost any facility. Due to the energy-intensive nature of data centers, this impact is magnified, and, consequently, careful attention must be made to the design of control systems to optimize and maintain energy-efficient operation. The focus of this chapter is on control of the air-conditioning and ventilation systems that serve datacom facilities. As such, the two sections of the "energy pie" (see Figure 6.1) that this chapter will impact are the

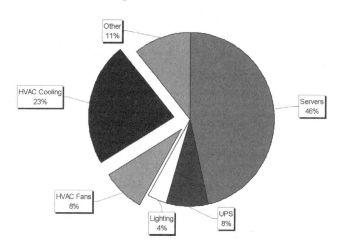

Average Data Center Power Allocation

Figure 6.1   Energy consumption impact of HVAC controls and energy management (LBNL average of 12 data centers).

"HVAC Cooling" and the "HVAC Fans" slices, which comprise an average of 31% of datacom facility energy consumption.

6.2   CONTROL SYSTEM ARCHITECTURE

This section looks at the development of the system architecture for the control system from the standpoint of functionality, energy efficiency, and reliability. A central goal is to discuss how reliable system control can be obtained in alignment with energy-efficient control strategies.

6.2.1   Background

Traditionally, building mechanical control systems are designed from a "performance specification" standpoint. Control strategies are generally developed as required to operate the mechanical plant both reliably (as a primary attribute) and efficiently (as a secondary attribute). In the past few decades, the control system architecture most typically employed is called distributed control. Distributed control employs a digital control system with a collection of control modules. All modules, each with its own subprocess specific function, are interconnected to carry out integrated data acquisition and control. An operator typically has control over most setpoints and other aspects of the control system through a supervisory computer. Failure of this supervisory computer system is not fatal to overall system operation, as distributed controllers have the ability to function independently.

6.2.2   Mechanical and Control System Reliability

Properly designed and distributed controls can achieve both reliability and energy efficiency. The main thrust of mechanical system reliability in a data center has been to ensure appropriate environmental conditions for datacom equipment on a 24-hour/7-day a week basis. Historically, this has been accomplished by providing redundant mechanical systems with little regard for building operating costs. With the current improvements in technology, today's control systems are capable of high-level control strategies that enable efficient performance of mechanical systems without compromising reliability. In general, data centers have been most at risk during "changes of state." For instance, many data centers have not utilized economizer cycle control because the changeover from mechanical cooling to free cooling and back was deemed to be high risk.

For many datacom owners and operators, reliability is still the primary driver. The challenge in designing a datacom control system, therefore, is to design energy-efficient control strategies that optimize energy performance without adding risk to the data center. The control system must operate to ensure continuous availability of cooling to the datacom environment. Within that framework, the control system is tasked to sequence or stage cooling equipment as necessary to support the environment. Ideally, the energy-efficient control system performs that staging based upon system efficiency as well as load. Variable-speed control points can also be monitored and adjusted to reduce energy consumption when possible.

A typical datacom environment is supported by the following three major types of mechanical systems:

1. Central mechanical plant
2. Datacom environmental equipment spaces
3. Support space environmental equipment

The central plant, consisting of cooling towers, pumps, and chillers, typically has the highest energy consumption of these three components. The plant operates 24 hours per day/7 days a week and, thus, offers the most potential for energy savings from control strategies. In the central plant, designing a control architecture that achieves the goals of high-speed response and reliable operation is often the most important means of achieving energy-efficient control.

Datacom environmental control equipment has traditionally been operated to maintain return conditions at the inlet to the computer room air conditioner (CRAC or CRAH unit). Typical return air temperatures between 64.4°F–80.6°F(18°C–27°C) resulted in supply temperatures from the air conditioners between 50°F and 60°F (10°C and 16°C). These temperatures have generally been found to be colder than what is necessary to maintain the inlet conditions to the datacom equipment between 64.4°F–80.6°F(18°C–27°C), the recommended temperature range of the guidelines (Class 1) outlined in *Thermal Guidelines for Data Processing Environments* (ASHRAE 2009). With proper design of the hot-aisle/cold-aisle environment, supply temperatures much closer to the ASHRAE (2009) thermal guideline range can be used. This permits either higher chilled-water or evaporating refrigerant temperatures or less supply airflow or, likely, both. Ultimately, operating the cooling system at the warmest possible temperature that can meet the cooling load offers the most opportunity for energy savings. Operating the CRAC units utilizing supply air control rather than return air control will therefore provide energy and operating cost reductions when supply air temperatures can be elevated toward the ASHRAE (2009) Class 1 temperature range. Humidity control must also be maintained, as will be discussed shortly.

The support space environmental equipment usually has the lowest energy footprint of the three mechanical systems listed above. Often these spaces consist of electrical equipment (such as transformers and UPSs) that requires continuous cooling and ventilation. Although this equipment often requires air conditioning, the acceptable space temperature in these rooms is often higher than the 64.4°F–80.6°F(18°C–27°C) preferred by the datacom equipment. In these instances, the spaces can be supplied with less cooling airflow per unit of heat rejection.

6.3   ENERGY-EFFICIENCY MEASURES

The intent of this section is to document control strategies that can significantly reduce energy use in datacom facilities. Topics include part-load operation, outdoor air control, demand-based resets, air-side economizer sequences, water-side economizer sequences, thermal storage systems, and humidity control.

6.3.1   Part-Load Operation

Part-load operation of a data center offers an opportunity to achieve substantial energy savings. The largest opportunity for energy savings is in reduced and more efficient operation of the compressors, either in a large chilled-water plant or in individual CRAC units. Figure 3.8, for instance, shows the increase in efficiency of a centrifugal chiller at part-load operation, as well as the fact that VFDs can significantly improve part-load efficiency relative to inlet guide vane control. When selecting chillers for datacom applications, particular attention should be paid to the most likely operating range of the chiller over the range of its lifetime. Considerations in this regard include performance with redundant chillers (such as five operating chillers, one being redundant, serving the load of four fully loaded chillers), and the fact that chillers may initially be operating at low capacity if they are sized for future growth in the facility.

The same strategy can be used in smaller rooms where compressorized CRAC units are deployed. Units with multiple compressors that can run unloaded or at variable capacities take advantage of very efficient operation at lightly loaded conditions. This allows systems that were sized for future loads to still operate and provide additional redundancy without penalizing (and actually improving) the operating efficiency of the system.

Variable-speed drives should also be considered for all fans and pumps. In general, the power input to a fan or pump is proportional to the cube of the speed of the device. So, a decrease of 10% in fan speed results in a 27% decrease in energy consumption, and a 20% decrease in speed results in a 49% decrease in energy consumption. Legacy fan systems that employed variable inlet vanes, or the practice of allowing pumps to ride up their pump curves, can be energy inefficient. Replacing these control components (vanes and throttling valves) with VFDs can significantly improve system control while saving energy.

6.3.2   Outdoor Air Control

Dedicated outdoor air systems have been installed in many datacom facilities to control space pressurization and humidity. Normally the internal humidity load in a datacom facility is quite low (human occupancy is low) and the sensible load is quite high. The most significant humidification load is generally infiltration. Therefore, if a dedicated outdoor air system can pressurize the space to prevent infiltration, all of the humidification and dehumidification requirements of the facility can be met by the outdoor air system. By maintaining the dew point of the outdoor air cooling coil below the dew point of the datacom environment, the primary cooling coils in the facility can operate dry (removing only sensible heat), and some of the inefficiency that is typically associated with trying to use these coils to both cool and dehumidify (such as the need to overcool and reheat to maintain relative humidity) can be eliminated. The advantages of dew-point control for datacom facilities were identified as early as 1988 (Conner 1988).

During periods with low ambient temperatures, it is also possible to design a system with an air-side economizer cycle. This is covered in more detail both in Chapter 4, "Economizer Cycles," and in Section 6.3.4 of this chapter.

6.3.3 Demand Based Resets

There are a number of demand-based reset control strategies that can be employed to achieve part-load energy savings. Two that are discussed in some detail, due to the unique design conditions of datacom facilities, are chilled-water reset and supply air pressure reset in a raised floor plenum.

Chilled-Water Reset

Chilled-water reset at low load is a HVAC control strategy for an office building, but it is not as likely to be utilized in datacom facilities if the chilled water coils are being used for humidity control as well as temperature control. If, however, there is an independent system for humidity control, it may well be possible to reset the chilled-water temperature of the facility to improve the efficiency of the chiller. If chilled-water reset is used, care must be made to ensure that adequate inlet air conditions are maintained to the datacom equipment (with particular attention to high density equipment) at the part-load condition.

Supply Air Pressure/Use of Variable-Speed Drives (VSDs) to Maintain Underfloor Pressure for CRAC Units

In both Chapters 3 and 5, the reader is cautioned on the use of VSDs in raised floor applications. This section will help the reader understand how to use variable-speed fan control to not only improve energy efficiency, but also improve system performance. VSDs can be used in a datacom air-handling system to reduce energy consumption, but several factors must be considered in order to achieve good performance. In a typical datacom vertical underfloor (VUF) environment with a raised floor supply air plenum and numerous CRAHs with chilled-water coils throughout the space, the system operates as a constant volume system with capacity controlled by a modulating chilled-water control valve. The control valve varies the chilled-water flow through the cooling coil in response to a temperature sensor either in the space or more typically in the return air. If the computer room air-conditioning system is constant-air-volume, what is the purpose of installing the VSDs? In many datacom facilities, all of the CRAHs operate continuously, even the redundant equipment. Without variable-speed control, the standby operating capacity can increase the airflow by 20% to 25% over design values. In most computer rooms with supply air floor plenums, it is common practice for technicians to temporarily remove floor tiles during IT equipment maintenance and installation. When the floor tiles are out of place, proper airflow will be disrupted and the underfloor static pressure reduced. VSDs, however, can automatically compensate for the reduced underfloor static by increasing airflow.

For VSDs to be considered for a VUF plenum application, a sequence of operation must be developed. There are several factors to consider in developing the sequence:

1. The lowest allowable airflow rate of the CRAH that will maintain acceptable temperatures throughout the raised floor area.
2. The minimum static pressure that must be maintained in the floor plenum to provide acceptable air distribution.
3. The effects of lower airflow on the cooling capacity of the CRAH. Turndown, static pressure control point, the number and location of static pressure sensors, the grouping of the CRAHs, and the effect of lower airflow on the system are several of the factors that must be considered.

One of the first considerations is turndown ratio, the ratio of allowable airflow to the maximum airflow. Redundancy is important in determining the turndown ratio. If one out of five CRAHs are redundant units, then all units operating at 80% airflow will circulate 100% of design airflow. Another factor is normal cooling load. The actual operating load is less than the design load in most installations. The difference between the operating load and design load can vary greatly, but 80% of design load is used here for illustration. If the normal load is 80% of the design load and 20% of the units are redundant units, the CRAH fans need to provide 64% of their design airflow to meet the normal cooling load. Before the minimum airflow can be set this low, the airflow patterns in the room should be simulated with a CFD computer program, and the performance of the floor tiles should analyzed. The CFD can provide both the airflow and the temperature information that can be expected through the computer room at various CRAH unit airflows.

The impact of airflow variation on the cooling coil is another important consideration. While lowering the fan speed will save fan energy, lowering the airflow across the cooling coil may increase the cooling coil's latent cooling (thus increasing chiller energy use). As the control system reduces the airflow, the supply air temperature should not be allowed to decrease. The controls should have feedback between the fan speed control and the cooling capacity control so that the space dew point remains unchanged. The control sequence also needs to include static pressure override, so that if the space temperature exceeds the maximum design temperature, the fan speed increases the airflow to maintain the required inlet air temperature.

6.3.4 Air-Side Economizer Sequences

Standard Air-Side Economizer Control Strategy

The use of air-side economizers (see Chapter 4) provides for the capability to eliminate vapor compression cooling during colder ambient conditions. The dampers in the air handler will switch to 100% outdoor air when the enthalpy of the outdoor air is less than the enthalpy of the return air (if the air does not need to be cooled to saturation, the comparison should be energy cost instead of enthalpy, as different conditioning processes may be used for outdoor and return air to handle humidity). A concern of a "standard" air-side economizer is that it can introduce very dry air into the data center. If a minimum humidity level is required in the data center, such as the 41.9°F dew point minimum recommended in *Thermal Guidelines for Data Processing Environments* (ASHRAE 2009), there may be excessive humidification energy costs during winter operation. The energy cost impact of humidification is site specific, and needs to be quantified on an annual basis before employing this design.

Wet-Bulb Economizer Control Strategy

When an adiabatic cooling air-side economizer is used, it will provide free humidification during cold ambient conditions using the heat generated within the data center.

During more moderate outdoor air temperatures, the direct evaporative cooling system will greatly extend the economizer cycle's number of hours of operation, especially in more arid climates (see Table 4.3). These are the ambient economizer hours when outdoor dry-bulb temperatures are above the building supply air setpoint, but the wet-bulb temperature is less than the room's required maximum dew-point temperature. In dry western climates, it is not unusual for 25% to 40% of the annual ambient bin hours to reside within this portion of the psychrometric chart. These are all hours when an air-side economizer, without evaporative cooling, would require both mechanical cooling and humidification.

Positioning of the cooling coil downstream of the adiabatic device will provide the lowest refrigeration system load. Care should be taken not to provide too high of a relative humidity to a datacom equipment cold aisle with this strategy, since the adiabatic device may increase the CRAH unit discharge temperature to near saturation.

A ridged media-type evaporative cooler should be used to ensure that nothing but water vapor is released into the supply airstream, as carryover of moisture would also result in carryover of any bacteria and other suspended and/or dissolved contaminants that might be present in the water. The quality of water, and the possible need for water treatment should also be considered when using this type of system. For more information on evaporative cooling and humidification, please review the "Evaporative Air Cooling Equipment" chapter of the *2004 ASHRAE Handbook—HVAC Systems and Equipment* (ASHRAE 2004d) and the "Evaporative Cooling" chapter of the *2007 ASHRAE Handbook—HVAC Applications* (ASHRAE 2007g).

6.3.5   Water-Side Economizer Sequences

Water-side economizers are often used in data centers to conserve energy. The three methods that are typically used are integrated economizers (both modular and integral with the mechanical plant) and parallel economizers (with the mechanical plant only). In some cases, water-side and air-side economizers are both employed in the same facility. The operational sequence listed below is specifically for water-side economizers implemented with a mechanical plant. A similar approach is available with modular CRAC units.

Integrated Economizers

Integrated economizers can be used to unload the chiller when conditions are not suitable for complete economizer operation. The water-side economizer works in series with the chiller, and acts to precool the return water to reduce the load on the chiller. Since the chillers can operate for many hours at low part-load conditions, careful analysis of the part-load conditions will help to achieve both reliable chiller operation and high efficiency. Chiller head pressure control is recommended for this process.

As wet-bulb conditions continue to drop, the chiller can be cycled off and the economizer can be operated independently to meet the facility load.

The integrated water-side economizer can provide for a substantial reduction in the annual cooling cost of a data center. Mechanical cooling plant design, control, and implementation strategies, however, are quite important. Excessive chiller cycling, for instance, should be avoided, as it may increase maintenance costs and decrease overall cooling system reliability.

Parallel Economizers

Parallel water-side economizers are used to replace a chiller when the outside air wet bulb allows for complete economizer operation. The parallel economizer can be started after the cooling tower lowers its condenser-water supply temperature enough to directly (or indirectly, through a heat exchanger) produce the design chilled-water temperatures for the plant.

During the condenser-water temperature ramp down, it is advisable that a head pressure control valve be used on the chiller to permit parallel operation of the required chillers while transitioning in and out of economizer cycle. Tower capacity and pump sizing need to be designed for this parallel operation.

It should be noted that integrated economizers save more energy than parallel economizers (due to increased hours of operation, albeit it at part load), and if compliance with the economizer section of ASHRAE Standard 90.1 is required, the integrated economizer must be employed as part of the economizer design (refer to Section 6.5.1.3 of ASHRAE Standard 90.1-2004 [ASHRAE 2004a]).

Combined Water- and Air-Side Economizers

Some designers have employed both air-side and water-side economizers in their datacom facility designs. While a technical description of this design is considered to be beyond the scope of this text, certain climatic conditions may make this an economical approach. For instance, climates with low ambient wet-bulb temperatures (i.e., desert environments) may be able to use integrated wet-bulb economizers for many of the warmer hours of the year, with a switch to the air-side (or air-side wet-bulb economizer) for the winter months. If the additional energy savings of a combined approach are small, reliability and TCO considerations may dictate the choice of one or the other system.

6.3.6   Thermal Storage

Thermal storage systems (TSSs) have been used for many decades to reduce demand costs and, in some cases, energy consumption. Typically, thermal storage has a diurnal cycle: energy is stored during periods of low utilization and/or low energy cost, and consumed during periods of high utilization and/or high energy cost. The best match is with facilities that have a significant swing in diurnal loads, and those for which the high load and high energy cost are coincident.

Often the main function of the TSS in the datacom facility is emergency backup of the refrigerant plant. The TSS may not have the capacity to allow the refrigerant plant to be totally shutdown for extended periods of time, but at less than design loads, operating the TSS instead of the refrigeration plant could reduce the demand charge. Once a demand limiting strategy is implemented, the control system can provide the data to validate and fine tune to the demand limiting operation. Whichever TSS technology is used, the data collection capability of the control system can be an important tool. Load profiles can be compared to energy consumption of the refrigerant plant equipment. Is the refrigerant plant operating at less than design capacity? Does the refrigerant plant need to start an additional chiller with all its attendant equipment to meet a short-term load? The data from the control system can help answer these questions.

Some of the different types of TSS are described below.

Chilled-Water TSS

TSSs using chilled-water storage tanks have been used in many industries to trim energy costs based on real-time pricing. One of the attributes of chilled-water storage is its ability to provide emergency backup cooling in the event of a chiller plant outage. However, selecting and finding a location for a tank large enough to provide for extended operation of a large datacom facility without the refrigeration equipment can be a challenge. A TCO analysis should be undertaken to determine whether there is an economic advantage to sizing the tanks larger than that needed for a short-term chiller plant outage.

Ice Storage TSS

Similar to chilled-water TSS, ice storage TSS can be used to trim energy costs based on real-time pricing. The ice storage TSS can also be used for emergency support of the data center in the event of a major mechanical plant malfunction.

An ice bank storage system requires low-temperature chillers and controls that need to be carefully selected for high reliability. The chillers used in ice bank applications typically also operate at a low thermodynamic efficiency. As such, they actually increase net energy consumption, though the ability to shift refrigeration operation from periods of high electrical rates to periods of lower electrical rates, combined with the ability to store more energy in a smaller area than with a water tank, could make this system attractive.

Phase-Change Storage TSS

There are also phase-change thermal storage systems that allow for compact thermal storage at a higher temperature, with an associated increase in chiller efficiency.

6.3.7 Humidity Control

As mentioned in Section 6.3.2, "Outdoor Air Control," due to low human occupancy, little humidity is internally generated in a data center. This provides an opportunity, for data centers of a significant size, to centrally control the humidity in an efficient manner. Dew-point control, rather than relative humidity control, is often the control variable of choice. The primary advantage of central humidity control is to avoid the simultaneous side-by-side humidification and dehumidification that can often be found in a poorly commissioned data center with multiple CRAHs or other cooling units providing this function. If dehumification (in addition to humidification) is handled centrally, another advantage is that the cooling coils "on the floor" of the datacom facility can run dry. This allows for chilled-water reset (if deemed appropriate at part-load operation) without increased relative humidity.

Humidity control costs will also be reduced, in the long run, with the use of high-quality humidity sensors. Maintenance of the proper control range will be a side benefit that is obviously of importance in a mission critical facility.

The maintenance of a proper dead band for humidity control is also important to energy efficiency. The current range recommended by ASHRAE (41.9°F dew point–60% RH and 59°F dew point at the datacom equipment inlets) should be wide enough to avoid having to humidify in one part of the data center and dehumidify in another, as long as sensors are kept properly calibrated. A low minimum relative humidity setpoint decreases operating costs by using less steam or other humidification source energy. Increasing the maximum RH setpoint similarly reduces costs by reducing latent cooling.

6.4   ENERGY-EFFICIENCY RECOMMENDATIONS/BEST PRACTICES

The control system and the control sequences are very important components of energy-efficient HVAC system design for datacom facilities. It is difficult to point out specific highlights for this chapter because the specific measures to be taken are dependent on the overall system design, which is a function of climate, type of air distribution system, etc. Still, the following control considerations are important components of almost all datacom facility control system designs:

- Utilize an economizer cycle to reduce mechanical refrigeration costs. The cooling plant energy costs will be cut substantially (10%–50% or more) in almost all climates if an economizer cycle is utilized. Be sure to consider the climate-specific risk associated with contamination when using air side economizers.

- The supply chilled-water and supply-air temperatures should be raised as high as possible in the datacom facility, while still providing the necessary cooling. The higher temperatures not only increase the thermodynamic efficiency of the mechanical refrigeration cycles, but also greatly increase economizer hours of operation in most climates.

- Central humidity control can eliminate one of the historical causes of wasteful energy use in data centers: adjacent CRAC units "fighting" each other to maintain tight temperature and humidity tolerances. Central dehumidification can also allow cooling coils on the floor to run dry, allowing for chilled-water reset at light loads without impacting relative humidity.

- Developing an efficient part-load operation sequence for the central cooling plant is extremely important due to the uncertainly of cooling loads in a datacom facility.

- It may be considered a subset of efficient part-load operation, but developing sequences that allow for the slowing down of HVAC equipment with VSDs without a negative impact on the cooling of spatially diverse datacom equipment loads is a major and important challenge in a datacom environment.

- On a site-specific basis, thermal storage can provide significant cost and/or energy savings for a facility. This is most applicable to facilities with a significant diurnal load variation.

7

Electrical
Distribution Equipment

7.1   INTRODUCTION

Electricity distribution is the physical means used to deliver electricity from one point to another, as shown in Figure 7.1.

A power plant generates alternating current (AC) electricity that is stepped-up by a transformer for transmission. The electricity will travel through electrical substations, transmission lines, distribution lines, and electricity meters before entering a service entrance. Once at the premise or building, the electricity is brought to the datacom equipment by various methods that will be covered in this chapter. According to GridWise^{TM}, "...nearly $500 billion of electric infrastructures must be added by 2020 to meet load growth projected by the Energy Information Agency." Although the load is growing, datacom equipment is only one component of the heat output. There is variation in the level of electrical distribution losses in a data center, depending on the equipment installed and the configuration of the distribution system. The main electrical distribution loss for the data centers documented in the LBNL study was an average 8% loss for uninterruptible power supply (UPS) equip-

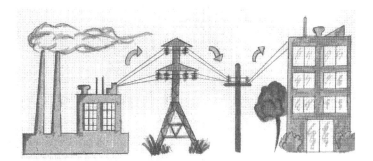

Figure 7.1   Electricity distribution.

ment. Electrical distribution losses, such as transformers, also made up some component of the 11% of "Other Losses" shown in Figure 1.4 (LBNL 2007a). Figure 7.2 shows (as offset pie slices) the areas of Figure 1.4 impacted by the components discussed in this chapter.

Another information source indicates that the electrical distribution and UPS losses represent 19% of the heat output contributed to a typical data center (APC 2003). Most datacom facilities probably fall into the 10%–20% range for these inefficiencies. This chapter will focus on electrical distribution from the building service entrance to the datacom equipment entry point in the data center, with emphasis on distribution techniques and energy-efficiency considerations.

7.2   DISTRIBUTION TECHNIQUES

A Federal Information Processing Standards publication, FIPS 94 (US DoC 1983), is one of the first documents that addresses the electrical environment and distribution to datacom equipment. FIPS 94 establishes the basic power distribution and the need for separation of various load types (e.g., lighting, air conditioning, IT) to minimize electrical disturbances. As the need for reliability, availability, and serviceability has increased, the power distribution arrangement has become more complex. Most electrical distribution paths to data centers include the utility service, main switchboard, alternate power sources, paralleling equipment,

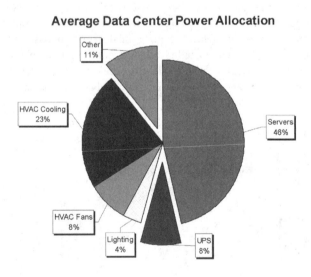

Average Data Center Power Allocation

Other 11%

HVAC Cooling 23%

Servers 46%

HVAC Fans 8%

Lighting 4%

UPS 8%

Figure 7.2   Energy consumption impact of electrical distribution equipment (LBNL average of 12 data centers).

and auxiliary conditioning equipment. Several electrical distribution schemes are available, including radial systems that do not duplicate components and require service interruption for maintenance, selective systems that include dual supply sources and possibly tie breakers, as well as loop systems that can receive power from either end of the loop via switches (DeDad 1997). Electrical distribution concepts and practices are constantly emerging to meet data center reliability, availability, and serviceability objectives. For example, a tri-isolated redundant design allows for three independent, operating power paths. Each power path supplies one-third of the data center load. However, if a single power path fails or is out of commission for maintenance, the datacom equipment load is transferred between the two remaining power paths (Yester 2006).

7.3 ENERGY-EFFICIENCY CONSIDERATIONS

A one-line diagram of the power path from the utility to the data center is shown in Figure 7.3.

Conventional UPS systems incorporate internal and external batteries that require a suitable temperature and humidity environment as well as have electrical losses from rectification of the utility power and charging circuitry. Flywheels are an alternative to batteries that should improve energy efficiency, but individual UPS product comparisons must be done to determine the overall benefit.

Figure 7.3 is only meant to be an example power path, as every installation is different in levels of redundancy, transfer capability, and protection from external as well as internal electrical disturbances. Because of the many distribution architectures, it is very difficult to quantify the efficiency of the total system. Figure 7.4 gives a simplified view of the power path with typical efficiencies.

The following sections will address the common building blocks, such as cabling, switchgear, transformers, and UPSs in detail.

7.3.1 Electrical Power Distribution Components

The electrical power distribution components, including, but not limited to, cables, switchgear, switchboards, and panelboards, have a heat output that is tied

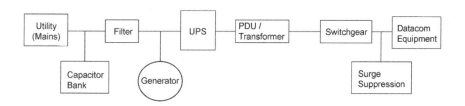

Figure 7.3 One-line diagram of the electrical distribution to the data center.

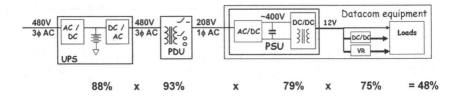

Figure 7.4 Power path with typical efficiencies.

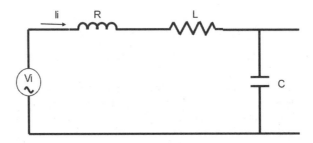

Figure 7.5 Equivalent AC circuit of a cable with one conductor.

directly to the load in the data center. Although equivalent impedance models, such as Figure 7.5, for distribution cables typically consist of resistance and inductance in series along with shunt capacitance, the capacitor has a very high capacitive reactance, X_C, and is nearly an open circuit at the 50 or 60 Hz fundamental frequencies used for AC distribution.

When current is flowing through cables, heat is generated per unit length of the conductor. The amount of heat loss, in watts, is calculated using the square of the current multiplied by the resistance, I^2R. The resistance is determined by the cross-sectional area of the conductor, stranding, and material (e.g., aluminum, copper). This information can be found in Chapter 9, Table 8, of the National Electrical Code (NFPA 2005). For example, a solid, uncoated 14 American Wire Gauge (AWG) copper conductor has a direct-current resistance of 3.07 ohms per 1000 ft (10.1 ohms per km) at 75°C. If 100 ft (30.5 m) of 14 AWG conductor is used to carry 10 amps, the I^2R loss equals 31 W.

The presence of harmonic distortion from nonlinear loads decreases the energy efficiency of the electrical distribution cabling. The reactive power is a product of the square of the harmonic current at a specific frequency and the inductive reactance, X_L. Since $X_L = j2\pi fL$ (where f is frequency and L is inductance), the inductive reactance is proportional to frequency for a given inductor value. Chapter 9, Table 9, of the National Electrical Code gives the inductive reactance for typical conductor sizes and also the alternating-current resistance, which is very close to the direct-current resistance. One study shows that by adding power factor correction, which is typically used for harmonics mitigation of nonlinear loads, the energy savings is in the range of 12%–21% for residential and commercial cable lengths of 40–100 ft. (12–30.5 m) (Fortenbury and Koomey 2006). The higher the power factor and efficiency of the equipment load, the lower the fundamental and high-frequency current in the electrical distribution components.

Rectangular busbars used in switchgear, switchboards, and panelboards also have a resistance that is based on the length, width, and height of a piece of metal as well as the material specific resistance, ρ. The equation used to calculate the resistance, R, for a busbar is

$$R = \rho \frac{1}{A}.$$

The parameter 1 is the length of the busbar, and A is the area or the height $\times$ width. For example, if the height $\times$ width of a busbar is 0.125 $\times$ 0.75 in. (0.3 $\times$ 1.9 cm), the area equals 0.094 in.^{2} (0.57 cm^2). If the length and specific resistance of copper busbar are 4 ft (120 cm) and 1.678 $\times$ 10^{-6} Ω-cm, the resistance is 353 $\mu\Omega$. With 200 A of current flowing in the busbar, which is the ampacity for the busbar used in the example, the I^2R loss is 14 W. The rectangular shape allows heat to dissipate efficiently due to the high surface area to cross-sectional area ratio. Overall, if the datacom equipment load is reduced in the data center, the I^2R losses of the cables, switchgear, switchboards, and panelboards will decrease.

7.3.2   Component Junction Points

Throughout the electrical distribution, there are many junction points in switchgear, switchboards, and panelboards that are connected together by mechanical fasteners to ensure a continuous path for current to flow. The mechanical fasteners must be tightened or torqued to ensure a good, solid connection. A loosely torqued connection can increase the contact resistance as the area of the conductive surfaces decrease. Although it is difficult to quantify the increase in resistance, the temperature does rise for a given current flowing through the junction point. Thermal infrared imaging can be used to detect the loose connections, as these manifest themselves by radiating energy in the form of heat.

7.3.3   Infrastructure Equipment

Besides the electrical power distribution components, there is infrastructure equipment, such as UPSs and transformers, that are used to condition and step down the AC voltage and current. There are three basic types of static UPS topologies: standby, line-interactive, and double conversion. Standby UPSs pass through utility power to the load with minimal filtering and are the lowest cost option. Line-interactive UPSs enhance the standby UPS with voltage regulation. Double conversion UPSs re-create the utility sine wave with an AC to DC rectifier and a DC to AC inverter. Double conversion UPSs offer the most power protection, but are typically the most expensive option.

Energy sources for UPS systems include internal and external batteries or flywheels. A flywheel consists of a contained rotating disc used to store energy. Energy is released during power disturbances or disruptions to provide for continuous power. The available runtime from a flywheel is close to constant and very predictable regardless of energy discharge frequency. Batteries and flywheels may be used in tandem, where the flywheel responds to the more frequent, short duration power disruptions that degrade batteries quickly, thus preserving their life to handle the less frequent, longer duration events. They may also be used independently. Battery life ratings are based on operation at 25°C, whereas flywheels can operate over a wider temperature range with no effect on life. This difference leads to energy savings on cooling required for the battery room. Both incur electrical losses from charging circuitry.

Although each UPS type behaves differently either in normal operation or while on battery, the general trend in the UPS industry is a smaller footprint, lower cost designs, higher efficiencies, higher performance, and more features. As shown in Figure 7.6 (CEMEP 2002), the trend in UPS technology from thyristors to insulated gate bipolar transistors (IGBT) has resulted in a 15% increase in efficiency.

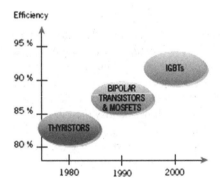

Figure 7.6 UPS semiconductor technology trend (Source: European Committee of Manufacturers of Electrical Machines and Power Electronics).

Even though the general trend shows efficiency increasing over time, certain topologies have higher efficiency. For example, line-interactive flywheel UPS efficiency is 98%, whereas double conversion UPS efficiency is 86%–95% at full load (Ton and Fortenbury 2005a). The efficiency of the UPS is not only based on the topology and technology, but also determined by the type of datacom equipment load, linear or nonlinear, and the amount of datacom equipment load. Figure 7.7 (Ton and Fortenbury 2005a) shows the impact that the type and amount of datacom equipment parameters have on UPS efficiency.

UPSs are available in single-phase and three-phase configurations. Three-phase UPSs are typically used to supply power and protect the datacom equipment load from disturbances at a data center level, whereas much smaller rated single-phase UPSs are typically installed at a rack or cabinet level. Balancing the datacom equipment load on the output of a three-phase UPS is important to minimize the potential for overloading the UPS and to increase the energy efficiency of the inverter, which takes the DC voltage from the rectifier and converts it to AC voltage for the load.

Transformers are used to step down the voltage to a value usable by the datacom equipment load. Even though the transformer is usually sized based on kVA and the AC voltage, there are efficiency parameters that can be controlled. Transformer heat load losses can be grouped into core losses, coil losses, and impedance. The core losses, also called the no-load losses, are a result of the magnetizing current that energizes the core. The amount of loss is constant and based on the kVA of the transformer as well as the core material and volume. Since transformers are typically loaded to a kVA that is much less than rating, selection of a transformer with capacity closer to the datacom equipment load can reduce core loss (NEMA 1996). The coil

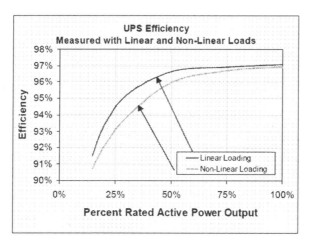

Figure 7.7 UPS efficiency as a function of equipment type and load (Source: Lawrence Berkeley National Laboratory).

losses, also called the *load losses*, are a result of the winding material and size that feed the connected equipment. The transformer load and no-load losses are not usually published or available on a nameplate, but temperature rise can be used as a rough indicator of transformer efficiency.

The standard temperature rise options for dry-type transformers are 176°F (80°C), 239°F (115°C), or 302°F (150°C), and, for liquid-filled transformers, are 131°F (55°C) or 149°F (65°C). The rise values are based on a maximum operating environment of 104°F (40°C) and rated capacity. This means the average winding temperature is 248°F (120°C) for a 176°F (80°C) transformer, and 374°F (190°C) for a 302°F (150°C) transformer. The operating energy of the 176°F (80°C) transformer is 13%–21% lower than the 302°F (150°C) transformer (CDA 2003).

The impedance of the transformer is on the nameplate, typically specified as a per unit (p.u.) or percentage of rated capacity. The percent impedance is mainly used for short-circuit current calculations, but the actual impedance can be determined by using the rated kVA and secondary voltage. As a simple example, if a three-phase transformer is rated 1000 kVA, 480 Vac secondary, and 6% line-to-line impedance, the rated full-load amperage output is

$$A = \frac{1000 \text{ kVA}}{480 \text{ Vac} \times 1.73} = 1204 \text{ amps}.$$

The 1204 amps is the current that flows in the secondary if it is short-circuited. The actual impedance equals the voltage drop in the transformer divided by the following rated full-load amperage output:

$$Z = \frac{6\% \times 480 \text{ Vac}}{1204 \text{ amps}} = 0.024 \ \Omega$$

If a transformer is selected with a smaller percent impedance, the actual impedance at the secondary will decrease, thereby reducing the heat loss of the transformer. It is important to note that reducing the impedance of the transformer can have a negative impact on fault current calculations. Fault current calculations predict the maximum amount of amps at a given point in the distribution and are limited by the transformer and cable impedances. As the impedances are reduced, the fault currents increase based on Ohms law. Improvements in energy efficiency by reducing impedance may require a change in fault interrupting devices, such as circuit breakers and fuses (Graves et al. 1985). If a circuit breaker or fuse is not rated to handle the fault current, it could result in improper operation or failure to clear the condition without damage.

Nonlinear loads have caused internal heating challenges to transformers in data center environments because of eddy current losses and hysteresis. K-rated transformers, based on an IEEE recommended practice (IEEE 1998), are overrated to compensate for the harmonic loading and associated heating. Although a K-rated transformer is a measure of thermal performance to harmonic emissions, the

K-factor is either accomplished by lowering the impedance, derating components, or a combination of both. In addition to K-rated transformers, there are many types of transformers, including, but not limited to, transformers designed specifically for nonlinear loads, phase shifting transformers, and ferro-resonant transformers that have advantages and disadvantages in certain applications.

Besides UPSs and transformers, there is other electrical infrastructure equipment, such as noise filters, capacitor banks, surge suppressors, and emergency standby equipment, such as motor generators. Although these types of equipment are not addressed specifically in this chapter, this equipment may have an impact on energy efficiency and should be reviewed individually if used in the electrical distribution.

7.4   IMPROVING ENERGY EFFICIENCY

Impedance is distributed throughout the premises in the electrical power distribution components and electrical infrastructure equipment leading to the datacom equipment load in the data center. One way to improve the heat loss from the electrical distribution is to reduce the impedance. While there are many components and pieces of infrastructure equipment that comprise the total impedance, improvements should be sought in the higher impedance areas, namely, the UPS, transformers, and long conductor feeds. If multiple transformers are used with 5%–8% impedance, decide if the transformers can be consolidated into fewer transformers with lower impedance in the range of 1%–3%. Long conductor feeds should be designed for the highest available premises distribution voltage. Long conductor feeds of smaller diameter gauge wire should be minimized to reduce the heat loss. Figure 7.8 shows an installation where a low voltage transformer is outside the data center. The transformer feeds a power panel that supplies datacom equipment loads via circuit breakers. The heat loss in the conductors is 1939 W.

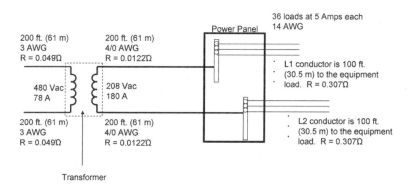

Figure 7.8   Simple electrical distribution to a data center.

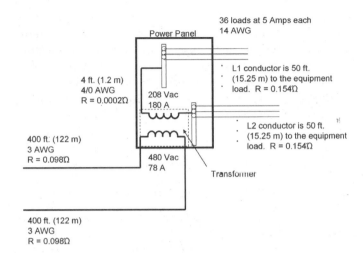

Figure 7.9 Simple electrical distribution to a data center with a lower resistance.

Figure 7.9 shows a slightly different setup where the transformer is located in the power panel, often called a power distribution unit (PDU). The PDU is also located in a more centralized spot within the data center, thus reducing conductor length to the datacom equipment. The heat loss for this configuration is 1483 W, approximately 25% less than that in Figure 7.8. Since Figures 7.8 and 7.9 use the same transformer and busbars, the transformer and busbar impedance are ignored for the calculations. Calculations for the above example are provided below.

Sup/Ret	Load(s)	I^2	R	Total (W)
2	36	25	0.307	552.6
2	1	32,400	0.0122	790.6
2	1	6084	0.049	596.2
				1939.4 (Sum)

Sup/Ret	Load(s)	I^2	R	Total (W)
2	36	25	0.154	277.2
2	1	32,400	0.0002	13.0
2	1	6084	0.098	1192.5
				1482.6 (Sum)

Figure 7.9 offers additional advantages with regard to the National Electrical Code. The neutral/ground bond is located close to the equipment and the voltage drop is better managed. If the transformer is considered a separately derived system, a neutral/ground bond is re-established. The presence of a solid neutral/ground bond close to the datacom equipment ensures that in the event of a line-to-ground fault, the fault current path is minimized and the overcurrent protection closest to the load is opened. National Electrical Code, Article 210.19(A), limits the voltage drop in branch circuit conductors to 3% (NFPA 2005). Excessive voltage drop can impair the starting and operation of lighting, heating, and motor loads as well as decrease their efficiency. As an example, if the applied voltage is below 10% of the rated voltage listed on an induction motor, the operating current would increase by 11%, the temperature would increase by 12%, and the torque reduced by 19% (NFPA 2005).

Also, if the equipment supports a rated voltage input of 480/277 Vac (typical input for North America datacom equipment is 100–127/200–240 Vac), it is possible to reduce the load loss from a step-down transformer as well as the heat loss from the cables. Operating at a higher voltage reduces the amperage a piece of datacom equipment requires based on Ohm's law. The lower amperage results in a significant reduction of I^2R loss even if the AWG has a greater resistance per unit length. It is important to note that 480/277 Vac is mainly used in larger electrical distributions for lighting loads, and there is a possibility that datacom equipment and lighting could share the same panels. The fluctuating nature of some datacom loads can cause visible voltage fluctuations and flicker.

Another way to increase the effectiveness of the electrical distribution is to increase the efficiency of the infrastructure equipment, especially with minimal loading. With the advent of dual power datacom equipment, electrical distributions are increasingly designed to support both of the datacom equipment power supplies with separate, independent electricity paths. There are multiple ways to accomplish dual electricity paths, but the longer the paths are separate, the longer the distance over which the electrical distribution is splitting the load and more electrical infrastructure equipment is operating at a very low load. Even though the cabling losses are less with dual electricity paths because the cables are carrying half of the normal datacom equipment amperage, it does not offset the losses in the electrical infrastructure equipment. As discussed previously, UPSs and other infrastructure equipment can be very inefficient with minimal loading. Data for the electrical infrastructure equipment must be requested, if not published, at all loading points to review the energy-efficiency curves (see Figure 7.6).

7.5 ENERGY-EFFICIENCY RECOMMENDATIONS/BEST PRACTICES

The electrical distribution requires careful planning to reduce impedances and, hence, heat load. The following considerations, at a minimum, should be taken into account:

- Choose UPSs that operate with high efficiency at light load and provide acceptable sensitivity to mains disturbances.
- Review the need for energy storage using batteries or flywheels, or a combination thereof, as individual configurations may represent savings on operational efficiency or total cost of ownership.
- Redundancy should be used only up to the required level. Additional redundancy comes with an efficiency penalty.
- Select transformers with a low impedance, but calculate the fault currents and install the proper over-current protection. Consider harmonic minimization transformers.
- Limit long conductor runs at the lowest datacom equipment utilization voltage by installing the PDUs as close to the rack load as possible.

8

Datacom Equipment Efficiency

8.1   INTRODUCTION

This chapter focuses on the largest component of energy usage within data centers and telecommunications facilities. As shown in Figure 8.1 and summarized by LBNL from a survey of 12 data centers, the largest component of energy usage in a data center is the datacom equipment. In fact, this is not unexpected and one would hope this was the largest component, as this is the component that generates the datacom function for business applications required.

Average Data Center Power Allocation

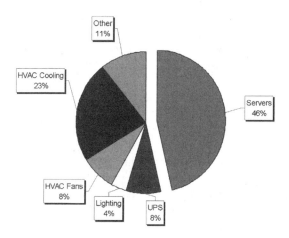

Figure 8.1   Energy consumption impact of servers (LBNL average of 12 data centers.)

As shown in Chapter 1, energy usage is a significant portion of the total facility operating costs of telecommunications facilities and data centers, and, therefore, energy efficiency is an important metric. Since datacom equipment is supplied power from systems of varying efficiency and requires cooling infrastructure to support the resulting heat load, improving the energy efficiency of datacom equipment is a critical step toward achieving an energy-efficient data center because the source heat load and the resulting HVAC system load is reduced.

It is estimated that approximately 40 TWh of electricity was used by servers and associated infrastructure equipment in 2005 in the United States alone, and worldwide usage was estimated at 120 TWh (Koomey 2007a). Continued growth in data center energy use is prompting governments and utilities to implement incentives to reduce datacom equipment energy usage. For example, the ENERGY STAR®️ Program (a joint program of the U.S. Environmental Protection Agency and the U.S. Department of Energy) includes specifications for workstations and desktop-derived servers in its computer program (ENERGY STAR 2007), and they supported the development of a server energy measurement protocol (Koomey et al. 2006). Servers are also included in the Electronic Product Environmental Assessment Tool (EPEAT©️), which is a system for identifying and verifying that products meet a set of voluntary environmental criteria, IEEE 1680–2006. The 80PLUS program enables buyers of datacom equipment to obtain rebates from participating utilities when they purchase equipment with a power supply that has been certified to meet specific efficiency requirements (80PLUS 2006).

Datacom equipment is continuing to become more energy efficient through the use of more efficient power-conditioning equipment, power supplies, lower power processors, and increasingly sophisticated power management.

Datacom equipment includes compute servers, network switches, and storage equipment. Many different types of servers are in use, from compact blade servers that may dissipate only about 200 W, to multiprocessor servers that can draw from 2 to 30 kW for some custom applications. Server applications vary from running weather simulations, considered high performance computing (HPC), to supporting Information Technology (IT) functions, such as e-mail, Web site hosting, and video streaming. In telecom applications, network switches provide call setup, billing, tracking, user authentication, etc.

Datacom equipment is generally housed in racks, in which many servers and/or switches may be combined. Typical rack dimensions are 24 in. wide by 2–4 ft deep by 7 ft high. Datacom equipment chassis are generally 17.5 in. wide in order to fit into the rack, the height is generally a multiple of 1.75 in. (1 U), and the depth is variable (EIA 1992). This chapter deals with energy efficiency as measured from the input of the rack to the loads.

Almost all loads within datacom equipment require direct current (DC) power at well-regulated low level voltages, in the range of 1 to 3.3 V. The input power to the rack is typically delivered as alternating current (AC) at voltage levels between

100 and 240 V—or as –48 V DC—but could be supplied at higher voltages. For example, power distribution at DC voltages in the range of 300–400 V has been proposed (Marquet et al. 2005), and some small-scale demonstrations have been built (Ton and Fortenbury 2006). The –48 V DC distribution voltage is from a battery bank, so it normally operates at battery float voltage of ~54 V, and may discharge down to 36 V. There are typically two or more stages of power conditioning between the rack input and the load, and losses are incurred in each stage. Losses are also incurred in distributing power within the rack and the equipment, especially where high power is delivered at low voltages.

This chapter introduces the components of datacom equipment and how each of these components affects energy efficiency. It also provides guidelines in selecting equipment to improve energy efficiency.

8.2   POWERING DATACOM EQUIPMENT

8.2.1   Load Requirements

While datacom equipment exhibits considerable variety, they have in common that almost all of the loads are based on active semiconductor devices, such as microprocessors or digital signal processors of varying complexity, volatile and nonvolatile memory devices, field programmable gate arrays (FPGAs), and other peripherals required to interface to the main components. All of the input power to these devices is converted to heat within the devices as they execute their respective functions. Exceptions to this are network switches with power over ethernet (PoE) outputs, in which some of the input energy exits the equipment on output cables capable of delivering both data signals and power at –48 V DC.

Most semiconductor devices within datacom equipment are manufactured using the complementary metal oxide silicon (CMOS) process. The power loss within the device has two main components: power loss due to leakage currents and power loss due to switching. The leakage power may be minimized by process enhancements. The switching loss is proportional to the frequency of switching and to the square of the supply voltage. Early versions of this process yielded devices for which 5 V could be used for the power rails. As the switching frequency increased, the voltage was reduced in order to limit the power loss, first to 3.3 V and then to progressively lower voltages. Today microprocessors operate at voltages close to 1 V. Some components still require a 5 V or 3.3 V rail voltage, and hard disk drives require 12 V for their internal motors; these voltages are referred to as *legacy voltages*.

Since the voltage required by a device is process dependent, and devices from many different generations of CMOS processes are found within the same equipment, many different voltages are required. In addition, devices are typically powered from separate rails when the equipment is in the "stand-by" mode, resulting in a requirement for additional voltage rails to be used in "stand-by." For example,

Table 8.1 lists the power rails required on a specific server, including three legacy voltages, six silicon voltages, and five standby voltages.

8.2.2   Power Delivery Network Design

The design of an appropriate power delivery network (PDN) to deliver all of the required voltages to the devices, while minimizing the losses incurred in the conversion and distribution of the electric power, is challenging. Many different approaches may be found to be implemented in different products, and this chapter is not meant as an exhaustive guide to all possible or available approaches, but, rather, will discuss the most typical configurations and general principles.

Some loads, e.g., microprocessors, require relatively high power, tens of watts to over a hundred watts, and delivering power to these loads at low voltage levels (1–5 V) results in load currents on the order of tens of amps to over a hundred amps. Conduction losses are proportional to the square of the current and to the resistance of the distribution path, which can be either a cable or printed wire board (PWB), and some high-end systems use busbars. The resistance can be minimized by either making the distribution path as short as possible, or by increasing the cross-sectional area of the distribution path, e.g., by using more layers of the PWB or to plate it with heavier copper. Each of these options have individual trade-offs

Table 8.1   Example—Voltages Required on Server Platform (Intel 2006)

Legacy voltages	12 V
	5 V
	3.3 V
Silicon voltages	~1 V (variable)
	1.8 V
	1.5 V
	1.2 V
	1.1 V
	1.05 V
Standby voltages	5 V
	3 V
	1.8 V
	1.5 V
	1.1 V

associated with cost. The load current is equal to the load power divided by the load voltage, which is determined by the silicon process, so the load current cannot be reduced. Maintaining a higher voltage for as long as possible and performing the final step down to the low level load voltage as close to the load as possible lowers the current and, therefore, the distribution losses over the greater part of the PDN.

Power-conditioning equipment consists of a combination of transistors and diodes, inductors and capacitors, along with transducers, controllers, and drivers for the transistors. The transistors may be used in either the linear mode, where it is always ON to a greater or lesser extent, or it may be switched rapidly between an OFF and a fully ON state. The first type of design is referred to as *linear*, and the latter as *switch mode*. Switch-mode equipment is more efficient and is therefore generally used at higher power levels. Linear regulators are less expensive, since their parts count is lower, but they are also significantly less efficient. However, due to their lower cost, they are still widely used at lower power levels.

Switch-mode regulators used in the final stage of conversion to regulate the load voltages are known as *voltage regulators* (VRs), or *point-of-load* (PoL) *converters*. Their basic structure and typical efficiencies are discussed in more detail below. VRs are typically located close to the loads, either on the motherboard or on daughter cards that connect to the motherboard, so as to minimize distribution losses at the low voltages.

The upstream part of the PDN generates the input voltage to the voltage regulators, and contains one or more stages of power conversion. The main output voltage of the upstream PDN is typically 12 V for equipment with power requirements below ~3 kW, and 48 V for higher power levels. "Legacy voltages," such as 5 V and 3.3 V, are also supplied. While most VRs run off the main 12 V rail, some VRs use the lower legacy voltages as input, especially at lower power levels, as shown in Figure 8.2. Linear regulators generally also run off the lowest voltage rail so as to minimize their power loss.

Several implementation configurations for the upstream PDN are possible, and while there is ongoing research into implementations that would improve overall energy efficiency (Lee et al. 2006; Dou et al. 2006), only the most typical ones are presented here, as shown in Figure 8.3. A power distribution unit (PDU) at the rack level provides protection for each server and provides load balancing across the three phases for an AC mains input. This may be followed by a Power Supply Unit (PSU) located within the datacom equipment itself, as shown in Figure 8.3 (a) and (b). In (a), the PSU contains a multi-output converter, so all the output voltages are generated within the PSU, and in (b), the PSU has a single output converter that generates the main power rail and external DC/DC converters generate the legacy rails.

The PSU may also be located in an equipment chassis, as shown in Figure 8.3 (c) and (d). This is typical for blade servers where the PSU is located in the blade chassis, and it outputs either 48 V or 12 V, which is distributed across a backplane to all the blade servers. The choice of backplane voltage is a trade-off between distri-

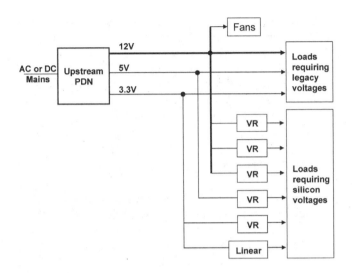

Figure 8.2 Power delivery to loads.

bution and conversion losses. For example, if the backplane voltage is 48 V, as in Figure 8.3 (d), distribution losses in the backplane are lower than at 12 V, but a converter is required within each blade to step the 48 V down to 12 V, whereas this conversion stage is not required with a 12 V backplane, as shown in Figure 8.3 (c). The PSU may also be located within the rack, as shown in Figure 8.3 (e). The typical output voltage here is ??–48 V because of the higher power. If the input power to the rack is DC mains at –48 V DC, no chassis level or rack level PSU is required.

The different levels of power conversion within the rack, linear regulators, voltage regulators, and PSUs are discussed in more detail below, and typical conversion efficiencies are provided.

8.3   POWER-CONDITIONING EQUIPMENT

8.3.1   Linear Regulators

Linear regulators are still used widely despite their low efficiency because of low cost. Their use is limited to lower power applications with low ratios of input to output voltage. A basic drawing of a linear regulator is shown in Figure 8.4.

The efficiency of a linear regulator, not accounting for drive losses, i.e. assume $I_{out} \cong I_{in}$, is

$$\eta = \frac{P_{out}}{P_{in}} = \frac{V_{out}I_{out}}{V_{in}I_{in}} \approx \frac{V_{out}I_{in}}{V_{in}I_{in}} = \frac{V_{out}}{V_{in}}.$$

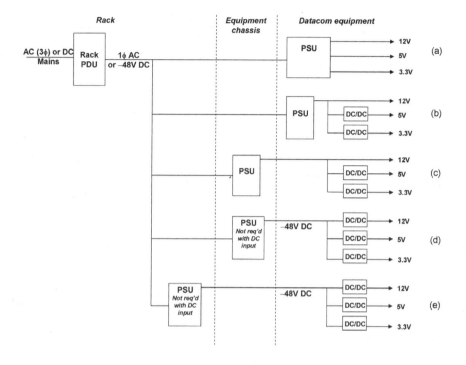

Figure 8.3 Power delivery network (PDN) implementations.

The efficiency is therefore proportional to the ratio between the input and output voltages, i.e. the step down ratio. For a linear regulator stepping down from 3.3V to 1.8V, the efficiency is only

$$\eta = \frac{V_{out}}{V_{in}} = \frac{1.8\,V}{3.3\,V} = 55\% .$$

8.3.2 Voltage Regulators/Point-of-Load Converters

Because of the low conversion efficiency of linear voltage regulators, switch-mode VRs have displaced them in almost all higher-power applications, such as powering the microprocessor and memory. VRs may be placed on the motherboard itself or may be plug-in modules. In many cases, OEMs design their own VRs from discrete components, but DC/DC converter modules are also commercially available. The most widely used topology for VRs is the buck converter, shown in Figure 8.5.

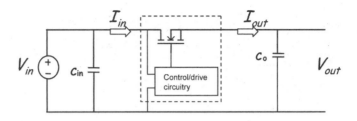

Figure 8.4 Linear regulator.

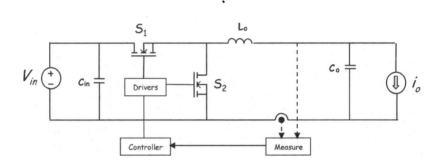

Figure 8.5 Switch-mode voltage regulator based on a buck converter topology.

The transistors S1 and S2 are turned ON and OFF in a complementary fashion, and the ratio of their ON times determines the output voltage. Losses are incurred every time a transistor is switched ON and OFF and, therefore, the switching losses are proportional to the switching frequency. However, the control bandwidth of the VR is also proportional to the switching frequency and the size of the output inductor decreases as the switching frequency increases, and, therefore, switching frequencies have steadily increased. Typical switching frequencies today are 100 kHz to 1 MHz, and some integrated VRs switch at several MHz. At these switching frequencies, metal oxide silicon field effect transistors (MOSFETs) outperform bipolar junction transistors (BJTs) and have become the components of choice.

In addition to switching losses, transistors also incur losses while they are on, since they appear as equivalent resistors to the circuit and there is ongoing development in reducing the on-resistance of the MOSFET as much as possible. Additional losses are due to conduction losses and core losses in the inductor and the losses in the driver circuitry for the MOSFETs and the controller. At the low output voltages required by the silicon loads, the conduction losses in the PWB are not insignificant.

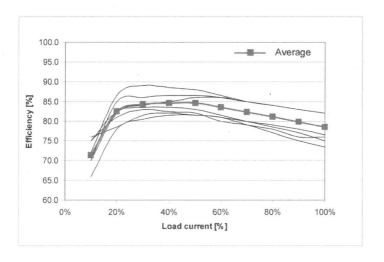

Figure 8.6 VR efficiency from 12 V input to 1.2 V output.

The efficiency of VRs vary widely, depending on the quality of components used (which translates to cost), the number of copper layers in the PWB, the input and output voltages, and the output current rating. Because losses in the VR add to the total heat load within the equipment, and heats up the PWB, VRs with higher power loads, such as for the microprocessor, have typically been designed to be more efficient. Typical efficiency curves for VRs of microprocessors are shown in Figure 8.6 as a function of load current, expressed as a percentage of the rated output current of the VR. The measured efficiency includes the conduction losses in the PWB, which constitute a significant portion of the total loss at heavy loads. Note that other VRs on the platform, for example, for the chipset and memory voltage rails, may not be as efficient.

It can be seen that the VR efficiency varies as a function of load. At heavy load conditions, the conduction losses dominate, since the currents are high and the conduction losses are proportional to the square of the current. Switching, controller and driver losses are independent of load, so at light load conditions, these losses constitute a larger percentage of the output power and, hence, the efficiency drops.

Figure 8.7 shows the measured efficiency of a VR as a function of load at different input and output voltages. For the most part, the smaller the ratio is between the input and output voltage, the higher the efficiency. For the same input voltage, the efficiency is higher for a higher output voltage. For the same output voltage, the efficiency generally increases as the input voltage decreases. However, full load efficiency drops more at very low input voltage, so an optimal VR input voltage can be

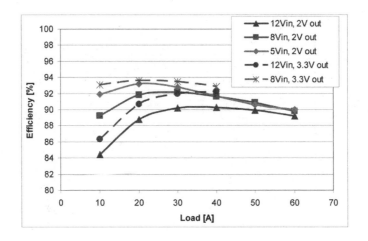

Figure 8.7   VR efficiency as a function of load at different input and output
voltages.

determined (Ren et al. 2003). Increased VR efficiency at lower input voltage has to
be balanced with distribution losses at the input, and the cost of providing a low
enough resistance in the distribution path to the input of the VR.

Where VRs are required to operate from a −48V input, a transformer-based
topology is typically used to better accommodate the step-down ratio. Figure 8.8
shows the efficiency of a 48 to 1.5 V VR, not including PWB losses, as a function
of load at different input voltages, ranging from 36 to 75 V DC.

8.3.3   DC/DC Converters

DC/DC converters are used to generate legacy voltage rails from the PSU
output. Typical configurations are a 12 or 48 V input and a 5 or 3.3 V output. The
technology is similar to that of a VR, with the buck converter widely used with 12 V
inputs, and transformer-based topologies used with a 48 V input.

The 48 to 12 V DC/DC converter is often referred to as a *bus converter*, since
it steps down from a 48 V bus to a 12 V bus. These bus converters may have a regulated, semiregulated, or unregulated output. Converters with semiregulated and
unregulated outputs are more efficient than converters with a regulated output voltage (Barry 2004). Figure 8.9 shows efficiency curves for 48 to 12 V bus converters
as a function of load and input voltage. The regulated bus converter provides a
12 V ± 5% output over an input voltage range of 36 to 75 V, while the unregulated
bus converter steps down the input voltage by a fixed ratio.

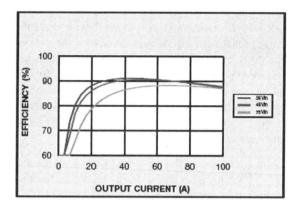

Figure 8.8 Efficiency of a 48 to 1.5 V VR as a function of load and input voltage (©2005 Artesyn Technologies).

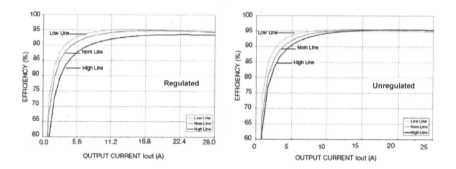

Figure 8.9 Efficiency of bus converters as a function of load and input voltage (©2006 Artesyn Technologies).

8.3.4 Power Supply Units

Many different power supply designs are found in industry, and only the most common are discussed here. As shown in Figure 8.3, PSUs receive either a single phase AC or a DC input and can provide single or multiple output voltages. Typical output voltages include a main power rail of 12 or 48 V, as well as other legacy voltages, the most common being 5 and 3.3 V, and the least common being −12V.

The main ingredients of a PSU with an AC input and single output voltage rail are shown in Figure 8.10. Most AC input power supplies are designed to accept voltages within the range of 90 to 264 Vac, also referred to as the *universal input range*. The low end of the range caters to Japan, with a nominal line-to-neutral voltage of 100 V, while the high end is a result of Europe, with a high nominal line-to-line voltage of 230 V, and the use of 240 V line-to-neutral in the United Kingdom. In the United States, typical line-to-line voltages are 208 and 240 V, and a line-to-neutral voltage of 120 V, which falls within this range. The main power conversion elements are an AC/DC converter and a DC/DC converter.

The AC/DC converter converts the AC input to an intermediate DC bus voltage, which is typically ~400 V. This is accomplished by rectifying the input voltage using a diode bridge, followed by a boost conversion stage that provides power factor correction (PFC), in order to limit the harmonic distortion of the input current drawn by the PSU. It is generally required that a PSU provide ride-through of a power outage for at least one power cycle, i.e., 16.7 ms in a 60 Hz system, and 20 ms in a 50 Hz system. To accomplish this, electrolytic capacitors are connected to the DC intermediate bus between the AC/DC and DC/DC converters.

The DC/DC converter steps down the intermediate DC bus voltage to the required output voltage, typically 12 or 48 V. The DC/DC converter also provides galvanic isolation between the input and output circuits. Similar to the VRs, controller and driver circuits are required, which adversely affect efficiency at light load conditions. In addition to the main converters, a standby power supply is also included with the PSU. The output rating of the standby power supply is significantly lower than the main converters. A small amount of additional loss is incurred in the electromagnetic interference (EMI) filter at the input of the PSU, which is required to meet EMI regulations. The PSU also has to power its own fan(s) required

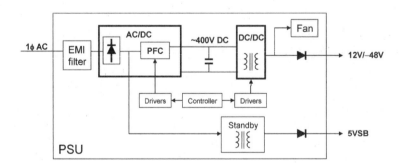

Figure 8.10 AC input PSU.

to cool its components, and the fans may also contribute to the cooling of the data-com equipment chassis.

Similar to the VR, the conversion efficiency is a function of load and input volt-age. The graph in Figure 8.11 shows measured efficiencies of a variety of power supplies as a function of load. The power supply power ratings range from 125 to 600 W, and six different form factors are included, some with multiple output rails and others with only a single 12 V output rail. These measurements were taken at an input voltage of 115 Vac, and Figure 8.12 shows how the efficiency of a specific PSU changes as a function of input voltage. It can be seen that, in order to maximize the PSU efficiency, it is desirable to supply it from the highest input voltage available within the input voltage rating of the PSU. For example, in a US installation, where three-phase 208 Vac is distributed to the rack, PSUs may be powered either from the line-to-line voltage of 208 Vac or the line-to-neutral voltage of 120 Vac. It is recom-mended to power from the line-to-line voltage to maximize efficiency.

A DC input PSU, as shown in Figure 8.13 for a single output, is similar to an AC input PSU, the main difference being that no AC/DC converter is required. Simi-lar to an AC PSU, the DC/DC converter may be a multi-output converter. The input to a DC input PSU is from a battery bank, so the PSU normally operates at the battery float voltage of 54 V, but it needs to operate down to the battery discharge voltage of 36 V. DC input PSUs are typically expected to provide ride-through for 10–12 ms, and, in order to minimize the required capacitance, the energy storage capacitors are charged to –72 V. The DC/DC converter within the PSU therefore has to operate over

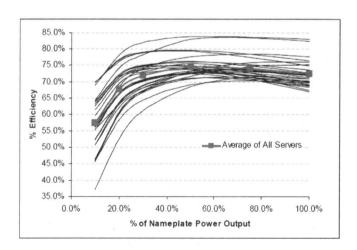

Figure 8.11 AC input PSU efficiency as a function of load (LBNL Interim Report 2005).

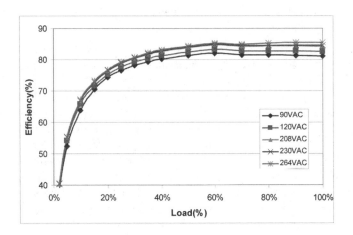

Figure 8.12 Effect of input voltage on PSU efficiency (Intel 2007).

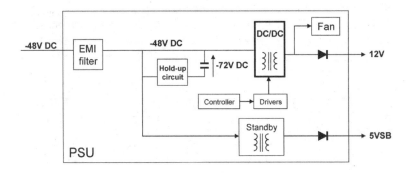

Figure 8.13 −48 V DC input PSU.

a voltage range of −36 to −72 V. The efficiency is typically higher than for an AC input PSU because no AC/DC conversion stage is required.

8.4   OTHER FACTORS AFFECTING ENERGY EFFICIENCY

There are many other factors and components within the datacom equipment that affect energy efficiency besides the efficiency of the power delivery equipment. These include the air-moving devices within the equipment, the ratio of the load to the equipment rating, the level of redundancy in the system, the level of power management employed, device technologies, and virtualization/consolidation. It is also worth noting that improved energy efficiency may also improve reliability.

8.4.1 Air-Moving Devices

The energy used by air-moving devices such as fans and blowers in IT equipment is another important and controllable energy consideration. Since almost all energy delivered to datacom equipment is converted to heat, fans or blowers are required to maintain components at safe operating temperatures. They may be located within the datacom equipment itself, which is typical for rack-mount servers, or in the equipment chassis, as for blade servers. They may also be within the rack, an option delivered by some OEMs. While some systems have AC-powered fans, most fans are powered by DC. Fan speed control is becoming more common, since lower fan speed can provide reduced acoustic noise, lower fan failure rate, and lower fan power (fan power is proportional to the cube of the fan speed). A concept for the future would be to coordinate building level fans and blowers and those in the equipment and control both from the on-board temperature sensors in the equipment.

8.4.2 Equipment Rating

As discussed earlier, conversion efficiency is a function of load, expressed as a percentage of the rating of the power conversion equipment. The conversion equipment has to be designed to handle the peak load power, but the load seldom operates at the peak load, since few applications require all of the loads to simultaneously operate at their peak power for extended periods of time. In addition, the PSU is rated for fully populated equipment; for example, a server may be designed to contain 2 processors, up to 16 memory cards, and up to 6 hard drives, but a particular platform may only have one processor, 4 memory slots, and 1 hard drive. This limits the peak power of this particular platform to ~60%–70% of the peak power of a fully populated platform. This means that during normal operation, the PSU will be operating at lower efficiency because of its light load. It is therefore recommended to use power supplies rated not much higher than the expected power draw from the platform; this is also referred to as *right-sizing* the power supply.

8.4.3 Power Management

Power management has two distinct roles in improving energy efficiency of data centers. The first is to scale power consumption to be proportional to its workload, i.e., the power is reduced when the workload is light, and the power is reduced to a minimum when the equipment is idle. This improves "performance per watt," and most modern servers have built-in technologies that address this requirement. Many power management techniques are borrowed from mobile platforms, e.g., laptops, where energy is conserved in order to extend battery life. Power management is most advanced in the processors, where its operating voltage and frequency is adjusted in order to minimize the required power to perform a specific task. Power management for other components is becoming more common over time. The achievable power savings varies considerably across platforms, depending on the level of power management technology available. It is recommended to activate power management features where offered in order to improve energy efficiency.

An emerging role for power management is to ensure that equipment power consumption is dynamically limited to a specified value. Many data centers contain racks that are only partially populated. The main reason for this is lack of knowledge of how much power a specific piece of datacom equipment consumes, and also a lack of control over the power consumed. Lack of such knowledge forces facility managers to overprovision power and cooling. Most of the time the result is that systems in a rack consume less than 50% of the power or cold air provisioned for the rack. The rack space as well as additional power and cooling capacity go unused. A secondary effect of this approach is that cooling systems and facility level power delivery equipment, including the UPS and PDU, operate at less than 50% of rating, which is a less-efficient operating point. This calls for new technologies that may allow a facility manager to allocate a power/cooling budget to a specific piece of equipment. The equipment then will ensure that its power consumption or heat load is lower than the allocated budget. This new approach will allow a facility manager to reduce the over allocation in power and cooling equipment, allowing them to operate more efficiently.

8.4.4   Redundancy

The level of redundancy built into a system also influences efficiency. For example, with 1 + 1 redundancy on the PSU, as shown in Figure 8.14, the PSU will always operate at a load <50%, even at peak loads, lowering its efficiency. This effect is compounded by oversizing of the PSU, which further lowers efficiency. In systems designed for very high reliability, redundancy is built in at all levels, the PSU, DCDC converters, and VRs, compounding the effect on overall efficiency.

8.4.5   Reliability

Improving the energy efficiency of datacom equipment results in less heat being generated within the datacom equipment itself, thereby reducing the temperature that components are exposed to. This, in turn, improves reliability, since the failure rate of electronic components increases with increased operating temperature.

8.4.6   Device Technologies

The implementation of multicore technologies has greatly improved the performance/power for processors. In addition, improvements in device structures that decrease the leakage current are being developed and will be put into production.

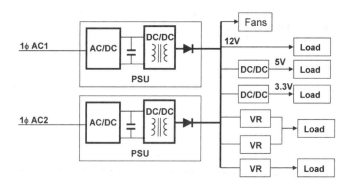

Figure 8.14 Example of redundant AC input power supplies and a redundant VR.

Therefore, replacing older equipment with more efficient equipment will improve the datacom facility efficiency.

8.4.7   Virtualization/Consolidation

Virtualization is an underutilized feature of datacom equipment in many data centers. It is a cost-effective and energy-efficient way to run two or more virtual computing environments, that is, running different operating systems and applications on the same physical hardware. It decouples the user from the physical hardware by providing a virtual system for the user. The physical hardware made up of processor, memory, and storage resources are divided into smaller granularities that allow multiple virtual machines to be operating simultaneously. Virtualization offers real benefits to datacom users in flexibility, power management, and scalability.

For those systems that are running at low utilization, some as low as 10% to 15%, virtualization allows the machine to be more effectively used by increasing this utilization and thereby increasing the performance/power efficiency of the machine. Virtualization provides a way to consolidate applications on a single shared physical system. The savings come in the amount of hardware required for the applications, reduced power since fewer machines are required to run the applications, and reduced cooling since fewer physical systems are required. Some overhead is required to implement the virtualization through hypervisors, but this is minimal compared to the gain that can be achieved through virtualization.

Virtualization enables the IT user to allocate resources where it is needed and when it is needed from a pool of virtualized systems.

8.5 ESTIMATING ENERGY EFFICIENCY OF DATACOM EQUIPMENT

Energy efficiency η for datacom equipment is defined as the ratio of the load energy E_{load} to the input energy E_{in}, i.e.,

$$\eta = \frac{E_{load}}{E_{in}} = \frac{\int p_{load}(t)dt}{\int p_{in}(t)dt},$$

where p_{load} is the load power, where the load includes microprocessors and other integrated circuits, volatile memory, and hard disk drives, and p_{in} is the input power delivered to the input of the PSU. If the load power can be broken up into a number of discrete power levels, this becomes

$$\eta = \frac{E_{load}}{E_{in}} = \frac{\sum_n (p_{load,n} \times t_n)}{\sum_n (p_{in,n} \times t_n)},$$

where $p_{load,n}$ and $p_{in,n}$ are the load power and input power respectively during the time period t_n.

As shown earlier, the efficiency of power-conditioning equipment varies as its load power varies relative to the rating of the power-conditioning equipment. As a result, even for the same equipment, the energy efficiency will vary as a function of the workload of the equipment. To illustrate this, the calculated power breakdown in a server under maximum and light load conditions is shown in Figure 8.15. Note that this breakdown is not representative of all servers; for example, some servers may contain more memory and less I/O, or a workload may be more processor or memory intensive. It can be seen that the power conversion losses constitute a much higher percentage of the total power under light load conditions, resulting in lower efficiency, than under heavy load conditions. This is important when it is taken into account that most datacom equipment operates at light load conditions for a significant portion of the time.

Datacom equipment exhibits great variety, with many different form factors, input voltages and power levels, power delivery, and cooling configurations available. In order to obtain the energy efficiency for a specific piece of equipment, the load for each component and the rating and load of each power conversion stage need to be known since the conversion efficiency is load-dependent. In addition, distribution losses in the printed wire boards (PWBs) and cables within the system need to be known. Such an exercise is beyond the scope of this work, so the simple example system shown in Figure 8.16 will be considered here to illustrate some key points. This system has four loads, with two of them operating off legacy voltages and the other two requiring a VR with efficiency η_{VR} to provide a well-regulated low

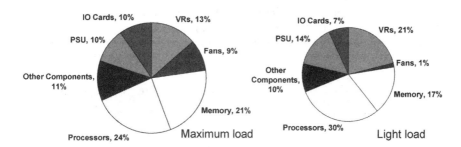

Figure 8.15 Breakdown of power dissipation within a server system at maximum load and at light load.

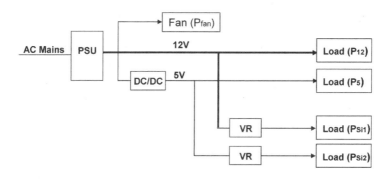

Figure 8.16 Example datacom power delivery example used to illustrate energy-efficiency calculations.

input voltage. The PSU with efficiency η_{PSU} has a single 12 V output rail, and the 5 V rail is generated by a DC/DC converter with efficiency $\eta_{DC/DC}$. A fan requires power P_{fan} to provide the necessary airflow.

At a specific load point, the load power can be calculated as

$$P_{load} = P_{12} + P_5 + P_{Si1} + P_{Si2},$$

and the input power is calculated as

$$P_{in} = \frac{1}{\eta_{PSU}}\left(P_{12} + \frac{P_{Si1}}{\eta_{VR}} + \frac{1}{\eta_{DC/DC}}\left(P_5 + \frac{P_{Si2}}{\eta_{VR}}\right) + P_{fan}\right).$$

Table 8.2   Calculated Energy Efficiency for Example System in Figure 8.16

	Units	High-Efficiency Power Conversion		Typical-Efficiency Power Conversion	
		Heavy Load	Light Load	Heavy Load	Light Load
P_Si1	W	100	60	100	60
P_Si2	W	50	30	50	30
VR eff	%	90	80	80	70
P_12	W	50	30	50	30
P_5	W	50	30	50	30
DC/DC eff	%	90	80	80	70
P_fan	W	20	15	20	15
PSU eff	%	85	75	75	65
Pload	W	250	150	250	150
Pin	W	351	273	448	361
System power delivery efficiency	%	71	55	56	42
Workload 1—weighted toward heavy load					
% time	%	70	30	70	30
Energy efficiency	%	66		52	
Workload 2—weighted toward light load					
% time	%	30	70	30	70
Energy efficiency	%	60		46	

System power delivery efficiency η at this load point can be calculated as

$$\eta = \frac{P_{load}}{P_{in}}.$$

Table 8.2 shows the calculated energy efficiency for this example system for two simple workloads, with typical- and with high-efficiency power-conversion equipment. The workloads are assumed to operate at a heavy-load condition $P_{load,h}$ for a period of time t_h and at a light-load condition $P_{load,l}$ for a time t_l. The two workloads are weighted toward heavy load and light load, respectively. The system's energy efficiency is then calculated as

$$\eta = \frac{P_{load,h} \times t_h + P_{load,l} \times t_l}{P_{in,h} \times t_h + P_{in,l} \times t_l},$$

where $P_{in,h}$ and $P_{in,l}$ are the input powers during the respective heavy- and light-load conditions.

As expected, energy efficiency for the workload weighted toward light load is lower than for the one weighted toward heavy load. The high-efficiency power delivery system also delivers higher energy efficiency, as expected.

In order to calculate the energy efficiency of the equipment as described above, the load for each component and the rating of each power conversion stage are required. Since this information is typically not available, a methodology has been proposed to represent datacom equipment efficiency as the ratio of the performance of the equipment to the input power of the equipment (Koomey et al. 2006). An energy-efficient piece of equipment will reduce the input power for a given performance or increase the performance achieved per unit of power delivered to the equipment. The initial protocol is focused on transaction-based benchmarks, where performance is measured in specific work units, such as Web pages served per second. The protocol calls for taking measurements at different levels of performance, including maximum, average, and idle workloads. A formal industry standard is expected to be released later in 2007 based on this protocol (SPEC 2007).

8.6   ENERGY-EFFICIENCY RECOMMENDATIONS/BEST PRACTICES

Improving the energy efficiency of datacom equipment is an important step toward achieving an energy-efficient data center. There is great variety in datacom equipment function, power level, power delivery architectures, power converters, and power supplies, which makes it hard to cite a typical efficiency for the equipment. This chapter discussed some of the more common architectures and conversion hardware, and presented a methodology for determining equipment energy efficiency.

In order to optimize the energy efficiency of datacom equipment, consider the following:

- Replace older equipment with more efficient designs.
- Install high-efficiency power supplies with power factor correction.
- Power equipment from the highest input voltage available within its input voltage rating range.
- Activate power management features where offered.
- Provide variable-speed fans or blowers and optimize their control and/or coordinate with building systems.
- Avoid using power supplies rated for much higher power than expected on the platform ("right-size").
- Use only the level of redundancy required to meet the availability requirements.
- Employ virtualization/consolidation.

9

Liquid Cooling

9.1   INTRODUCTION

Most current large data center designs use chilled air from air-conditioning units (modular or centralized) to cool datacom equipment. With rack heat loads steadily increasing, many data centers are experiencing difficulty in cost-effectively meeting datacom equipment's required inlet air temperatures and airflow rates. Therefore, more and more facilities are implementing liquid cooling, and others are considering it. While the initial rationale for liquid cooling has been to improve cooling and compaction, liquid cooling can also reduce the energy consumption of the "HVAC Cooling" and "HVAC Fans" components shown in Figure 9.1. Together, these two slices represent 31% of the energy consumption in an "average" datacom facility.

Liquid cooling is defined as "the case where liquid must be circulated to and from the entity for operation" (ASHRAE 2006c). For instance, a liquid-cooled rack/cabinet defines the case where liquid must be circulated to and from the rack or cabinet for operation. This definition can be expanded to liquid-cooled datacom equipment and liquid-cooled electronics. Liquid-cooled applications can deploy open or closed cooling architectures, or single- or two-phase cooling.

Open versus closed architecture indicates whether electronics in the cabinet are exposed to the room air (open) or if the cabinet is closed and the cooling air is continuously circulating within the cabinet. An example of an open system is a rack with a back door heat exchanger. Cool room air enters the front of the cabinet, absorbs heat as it passes over the electronics, then passes through the heat exchanger in the rear door before it exits. In this scenario the liquid is removing some or all of the heat from the cabinet and relieving the load on the room air-conditioning system. In a closed design, air is circulated through the electronics, passes through an air-to-liquid heat exchanger located in the cabinet, and is then returned back to the electronics. In this case the liquid in the heat exchanger absorbs all of the heat, and the recycled air meets the incoming air temperature specifications for the datacom equipment. The advantages and disadvantages of each type of system are detailed in Table 9.1.

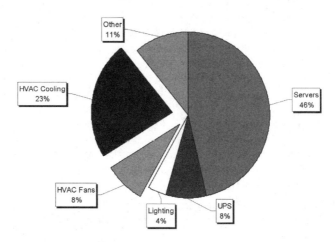

Average Data Center Power Allocation

Figure 9.1 Energy consumption impact of liquid cooling (LBNL average of 12 data centers).

Table 9.1   A Comparison of Open versus Closed Architecture for Liquid Cooling

Open Architecture System	Closed Architecture System
Advantages	**Advantages**
1. System Redundancy	1. Low audible noise
2. First cost and operating cost	2. Deployable as a single rack
3. Uses room as heat sink for emergency mode	3. Does not require hot-aisle/cold-aisle room layout
4. Does not limit rack selection	
Disadvantages	**Disadvantages**
1. Audible noise for some designs	1. Requires server OEM certification
2. Room solution	2. Short ride-though on cooling loss for some designs

Single-phase versus two-phase refers to the liquid coolant state. If the coolant enters and exits the system as a liquid, the system is considered single-phase. If the coolant enters as a liquid and exits as a gas or gas/liquid combination, the system is considered two-phase. The most common fluids used in single-phase systems are water and ethylene glycol/water (EGW). Water is low cost but it has electrical hazard concerns. EGW has poor thermal characteristics and electrical hazard concerns, but provides corrosion protection and resists freezing when run outdoors in cold climates. The most common fluids used in two-phase systems are refrigerants (R-134a, carbon dioxide [R-744]) and dielectrics (e.g., FC-72 or OS-110).

More information on liquid cooling definitions, configurations, and guidelines can be found in ASHRAE (2006c).

This chapter covers energy efficiency and total cost of ownership of liquid-cooled systems. The general industry consensus is that liquid cooling technology provides a significant improvement in energy efficiency. The chapter includes the following:

- A comparison of key liquid coolant properties (Section 9.2)
- A discussion of energy optimization with specific liquid cooling fluids (Section 9.3)
- A discussion of rack-level liquid cooling (Section 9.4)
- A review of TCO aspects of liquid cooling other than energy (Section 9.5)
- An example comparison showing typical pumping energy requirements for three fluids (Section 9.6)
- A summary with recommendations (Section 9.7)

9.2   COMPARISON OF KEY LIQUID COOLANT PROPERTIES

Tables 9.2a and 9.2b list the key coolant properties for a number of coolants relatively widely used in the electronics cooling industry.

For single-phase cooling, the important properties include the thermal conductivity, specific heat (at constant pressure), density, and viscosity. The thermal conductivity plays a critical role in conducting heat from any heat source into the coolant. The specific heat plays an important role in the coolant's ability to store any absorbed heat. The viscosity has an impact on the required pumping power when moving the coolant.

For two-phase cooling, the important properties include the properties named for single-phase cooling, in addition to the latent heat of vaporization and surface tension. The properties named for single-phase cooling play a role because the heat transfer is, in effect, single-phase heat transfer up to the point that the coolant reaches saturation. Upon reaching saturation, any additional heat absorbed by the coolant goes to transforming saturated liquid into saturated vapor. The key property involved in the transition from saturated liquid to saturated vapor is the latent heat of vaporization. This transition from liquid to vapor occurs at a fixed temperature. The

Table 9.2a Comparison of Key Coolant Properties (I-P Units)

Coolant	Freezing Point, °F	Viscosity, lb/ft·s	Thermal Conductivity, Btu/h·ft·°F	Specific Heat, Btu/lb·°F	Density, lb/ft³	Latent Heat of Vaporization, Btu/lb	Surface Tension, lbf/in.
Dielectrics							
FC-87 (PF5050)	−175	3.23E−04	0.033	0.251	103.6	44	5.63E−05
FC-72 (PF5060)	−130	4.64E−04	0.034	0.251	105.3	38	6.23E−05
HFE 7000	−184	3.24E−04	0.044	0.308	88.3	63	
HFE 7100	−211	4.16E−04	0.040	0.282	95.4	52	
Aromatic (DEB)	<−112	6.72E−05	0.081	0.408	53.7		
Aliphatic (PAO)	<−58	6.05E−04	0.079	0.516	48.0		
Silicone (Syltherm XLT)	<−166	9.41E−04	0.064	0.384	53.0		
Water and Water-Based							
Water	32	7.26E−04	0.347	1.004	62.3	1058	4.15E−04
Ethylene glycol/water (50:50 v/v)	−36	2.55E−03	0.215	0.788	67.8		
Propylene glycol/water (50/50 v/v)	−31	4.30E−03	0.209	0.816	66.3		
Methanol/Water (40:60 wt./wt.)	−40	1.34E−03	0.232	0.854	58.3		
Ethanol/Water (44:56 wt./wt.)	−26	2.02E−03	0.220	0.840	57.8		
Potassium Formate/Water (40:60 wt./wt.)	−31	1.48E−03	0.307	0.768	78.0		
Refrigerants							
R-134A	−154	1.40E−04	0.048	0.337	76.4	93	5.00E−05
R-410A	–	8.80E−05	0.060	0.398	67.6	101	3.66E−05
R-744 (carbon dioxide)*	−70	4.95E−05	0.049	0.815	48.4	66	2.86E−05

* R-744's flash point (i.e., boiling point at atmospheric pressure) is lower than its freezing point. Therefore, it goes direct from solid to gas (sublimation) under these conditions.

Table 9.2b   Comparison of Key Coolant Properties (SI Units)

Coolant	Freezing Point, °C	Viscosity, kg/m·s	Thermal Conductivity, W/m·K	Specific Heat, J/kg·K	Density, kg/m³	Latent Heat of Vaporization, kJ/kg	Surface Tension, dynes/cm
Dielectrics							
FC-87 (PF5050)	–115	4.81E-04	0.056	1045	1660	103	9.9
FC-72 (PF5060)	–90	6.90E-04	0.058	1045	1688	88	10.9
HFE 7000	–120	4.83E-04	0.076	1285	1415	148	
HFE 7100	–135	6.20E-04	0.070	1173	1529	121	
Aromatic (DEB)	<–80	1.00E-04	0.140	1700	860		
Aliphatic (PAO)	<–50	9.00E-04	0.137	2150	770		
Silicone (Syltherm XLT)	<–110	1.40E-03	0.110	1600	850		
Water and Water-Based							
Water	0	1.08E-03	0.598	4182	998	2461	72.7
Ethylene glycol/water (50:50 v/v)	–37.8	3.80E-03	0.370	3285	1087		
Propylene glycol/water (50/50 v/v)	–35	6.40E-03	0.360	3400	1062		
Methanol/Water (40:60 wt./wt.)	–40	2.00E-03	0.400	3560	935		
Ethanol/Water (44:56 wt./wt.)	–32	3.00E-03	0.380	3500	927		
Potassium Formate/Water (40:60 wt./wt.)	–35	2.20E-03	0.530	3200	1250		
Refrigerants							
R-134A	–103.3	2.09E-04	0.083	1403	1225	217	8.75
R-410A	—	1.31E-04	0.103	1658	1083	234	6.41
R-744 (carbon dioxide)*	–56.5	7.37E-05	0.084	3396	775	153	5

* R-744's flash point (i.e., boiling point at atmospheric pressure) is lower than its freezing point. Therefore, it goes direct from solid to gas (sublimation) under these conditions.

surface tension is a property that is responsible for causing liquid to form droplets on the surface being cooled. The lower the surface tension, the greater the tendency for the liquid to spread over a surface and effectively wet it. During two-phase cooling it is important for a surface to remain wet, so coolants with lower surface tensions are preferred.

Table 9.3 lists the same properties shown in Tables 9.2a and 9.2b, but ratioed to the corresponding value for water. The reason for choosing water as a baseline is simply that the properties of water are well known to most people, and, thus, the normalized values can be understood quickly.

9.3   ENERGY OPTIMIZATION WITH SELECTED LIQUID COOLANTS

There is a significant variation in the properties of different types of liquid coolants, and, as such, it can be difficult to make an "apples-to-apples" comparison for a specific application. Without trying to indicate that one fluid is better than another, the intent of this section is to look at some representative classes of liquid coolants, and provide guidance on energy optimization within each of these classes. For those interested in the physical properties of the liquids discussed in this section, please refer to Tables 9.2 and 9.3.

9.3.1   Water-Based Liquid Cooling Systems

Water-cooled (or glycol-cooled) systems have a number of important parameters that can be optimized for energy efficiency. Most energy consumption in water-cooled systems can be placed into four broad categories. These categories are:

1. Pumping Power
2. Fan Energy (not specifically to move the water, but to drive the air side of any air-to-liquid heat exchangers)
3. Chilled Water Production Efficiency
4. Heat Rejection

A discussion of each of these categories, and factors that can increase energy efficiency, follows.

Pumping Power

Pumping power is dependent on many parameters, including the following:

*   System piping design (primary, primary/secondary, etc.)
*   Water-side differential temperature
*   Pipe sizes
*   Heat exchanger pressure drops
*   Fluid viscosity
*   Fluid specific heat capacity
*   Use of VFDs

Table 9.3   Ratios of Key Coolant Properties to a Baseline of Water

Coolant	Viscosity, kg/m·s	Thermal Conductivity, W/m·K	Specific Heat, J/kg·K	Density, kg/m³	Latent Heat of Vaporization, kJ/kg	Surface Tension, dynes/cm
Dielectrics						
FC-87 (PF5050)	0.45	0.09	0.25	1.66	0.04	0.14
FC-72 (PF5060)	0.64	0.10	0.25	1.69	0.04	0.15
HFE 7000	0.45	0.13	0.31	1.42	0.06	
HFE 7100	0.57	0.12	0.28	1.53	0.05	
Aromatic (DEB)	0.09	0.23	0.41	0.86		
Aliphatic (PAO)	0.83	0.23	0.51	0.77		
Silicone (Syltherm XLT)	1.30	0.18	0.38	0.85		
Water and Water-Based						
Water	1.00	1.00	1.00	1.00	1.00	1.00
Ethylene glycol/water (50:50 v/v)	3.52	0.62	0.79	1.09		
Propylene glycol/water (50/50 v/v)	5.93	0.60	0.81	1.06		
Methanol/Water (40:60 wt./wt.)	1.85	0.67	0.85	0.94		
Ethanol/Water (44:56 wt./wt.)	2.78	0.64	0.84	0.93		
Potassium Formate/Water (40:60 wt./wt.)	2.04	0.89	0.77	1.25		
Refrigerants						
R-134A	0.19	0.14	0.34	1.23	0.09	0.12
R-410A	0.12	0.17	0.40	1.08	0.10	0.09
R-744 (carbon dioxide)	0.07	0.14	0.81	0.78	0.06	0.07

System Piping Design. There has been considerable research over the years with different chilled-water pumping system designs to maximize controllability, while minimizing pumping energy (ASHRAE 2004b). Constant-volume systems were typically replaced with constant-volume primary/variable-volume secondary pumping systems. More recently, variable-volume primary-only systems have been utilized, now that chiller evaporators capable of handling variable flow have been developed. There is typically no simple answer to system design optimization: several system designs should be analyzed to determine the best system for the specific application.

Water-Side Differential Temperature. Everything else being equal, doubling the allowable differential temperature of a single-phase process will halve the required flow rate and reduce the required pumping power by 50%. Unfortunately, "everything else" is not typically equal, and the effect of lower flow rates and higher differential temperatures on heat exchanger performance and chiller efficiency must also be examined to optimize this variable. In some cases, increasing the differential may drop the supply water temperature below the dew point in the space, and the latent coil load will increase.

Pipe Sizes. for a given pipe diameter, the pressure drop in a piping system increases as velocity increases. Increasing pipe sizes will increase capital costs, but it also decreases pressure drop and allows pumps with a lower head to be selected, or VFDs to be run at a lower frequency to achieve operational cost savings.

Heat Exchanger Pressure Drops. There are several different types of heat exchangers in water-cooled systems. These include liquid-to-air heat exchangers, chiller evaporators (or condensers), and liquid-to-liquid heat exchangers, such as plate-and-frame heat exchangers used to isolate condenser water and process chilled water during water-side economizer operation. The selection of a heat exchanger typically involves a trade-off between first cost and performance, or even between water-side pressure drop and air-side pressure drop. Several selections should be made to determine the trade-offs involved and to make the best selection.

Fluid Viscosity. Higher fluid viscosity increases pressure drops in both heat exchangers and piping, so all things being equal, fluids with lower viscosities will have lower pump energy consumption. With water-based fluids, lower viscosities typically correspond to higher operating temperatures and/or lower percentages of glycol.

Fluid Specific Heat Capacity. Most water-based fluids have a specific heat capacity that is less than water. This means that the mass flow rate of the fluid must increase for a given amount of heat rejection, which typically increases pumping power. A full analysis of the energy impact of adding glycol must include the combined effects of (typically) higher density, lower specific heat, and higher viscosity compared to water. Most manufacturers provide basic technical information on glycols at various concentrations to allow for these determinations. One important maintenance note: for corrosion prevention, most manufacturers sell glycol with inhibitors and recommend that the percentage of glycol be maintained above a minimum level, such as 20%, for effective protection.

Variable-Frequency Drives (VFDs). VFDs can be used both to set the maximum pump flow and head during the commissioning phase of a project, and to offer operational savings during periods of low utilization once a facility is operational. The degree of savings will depend on the utilization schedule of the facility. A facility with staged loading will also benefit from VFD pump control, as the pumps can operate at low speed until the IT equipment is installed and operational. A review of fan laws (ASHRAE 2004c) and pump laws shows that the power input to both fans and pumps is proportional to the cube of the speed of the devices. As such, a fan or pump operating at 50% of its design speed requires only 12.5% of the theoretical power of the same device operating at full speed. The energy-saving potential of variable-speed drives thus becomes quite apparent when a device can operate at part load.

Fan Energy

Fans are known to consume a significant portion of energy in most data centers. Installation of liquid cooling usually reduces this percentage, but in many applications there are still air-to-liquid heat exchangers, and it is important to consider the design and location of these heat exchangers to minimize fan energy consumption. In an equipment cabinet, for instance, if the heat exchanger is located directly in the path of the equipment's hot exhaust airflow, then the heat can probably be transferred to the liquid with little or no additional fans needed. If the heat exchanger is not directly within the hot exhaust path, however, additional fan power is needed to redirect the air to flow across the heat exchanger. This additional fan power will also increase the room's acoustical noise level in open architecture systems, as described in Section 9.1.

Chilled-Water Production

The efficiency of chilled-water production is a function of many parameters, and these are described in Chapter 3 of this book. The most significant parameters are as follows:

- Chilled-water design temperature
- Condenser-water design temperature
- Process fluid temperature
- Chiller efficiency
- Number of hours of economizer operation

Once design conditions are known, several manufacturers should ideally be contacted to determine the efficiency of their chillers. In most cases, multiple selections will be available, and a trade-off will need to be made between first cost and energy efficiency. Different refrigerants will also typically have different efficiencies, and some may be better suited to the specific operating temperature range of a facility than others. Higher temperatures may preclude the use of certain refrigerants due to the higher operating pressure of the refrigerants.

Chilled-Water Design Temperature. The chilled-water design temperature may not be a variable for all installations, but to the extent that it is a higher temperature, it will typically increase both chiller efficiency and the number of hours of economizer operation.

Condenser-Water Design Temperature. Lower condenser-water temperatures almost always results in an increase in chiller efficiency.

Process Fluid Temperature. In many cooling system designs, the chilled water cools a process fluid, as shown in Figure 9.2. Higher process fluid temperatures will, in turn, allow for higher chilled-water temperatures, with the savings as indicated above. It is recommended that the process fluid temperature be maintained above the room dew-point temperature to eliminate the risk of condensation on the piping, coils, and server racks and hardware.

Economizer Operation. Economizer operation typically allows a water-side economizer to either (1) allow a chiller to operate with a lower utilization or (2) allow the chiller to be turned off completely. Computer analysis programs that calculate annual energy consumption can typically be used to estimate the energy savings available from economizer operation. The savings are usually substantial and dependent on all aspects of the system design, as well as local climatic conditions. Chapters 3 and 4 provide more quantitative information on savings opportunities.

Heat Rejection Efficiency

Several forms of heat rejection are typically available, and the cost and energy efficiency of these different forms of heat rejection can vary considerably. Types of heat rejection equipment include the following:

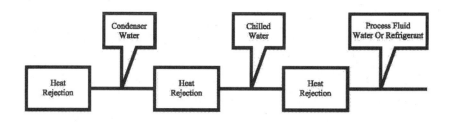

Figure 9.2 Heat rejection pathway for a common datacom facility.

- Open-cell cooling towers
- Dry coolers
- Evaporatively cooled heat rejection equipment
- Other options

Open-Cell Cooling Towers. Open-cell cooling towers are widely used in areas with a readily available and inexpensive source of makeup water. They are typically quite efficient and capable of rejecting heat near the ambient wet-bulb temperature. Towers can be selected to minimize fan energy and/or decrease approach temperature, but there is typically a first cost penalty for this increase in energy efficiency.

Dry Coolers. Dry coolers are often used with small computer-room air-conditioning units to provide for low-maintenance heat rejection. Unfortunately, the heat rejection efficiency is usually low due to the fact that condensing temperature is pegged to dry-bulb temperatures rather than wet-bulb temperatures, and the fact that increased fan energy is usually needed relative to open cell towers per unit of heat rejection.

Evaporatively Cooled Heat Rejection Equipment. The efficiency of evaporatively cooled equipment typically falls somewhere between the efficiency of open-celled towers and dry coolers. The ambient side of the heat rejection coil has considerable air blowing over it, but also water circulated to reduce the temperature of the ambient air to a point intermediate between the ambient dry-bulb and wet-bulb temperature.

Other Options. Other forms of heat rejection may also be possible, such as ground-source heat rejection, or rejection into a lake or river, and these can be quite efficient due to a lower heat rejection temperature. There may be severe environmental restrictions placed on the use of these heat sinks, however, so they should be considered only after appropriate research, discussions with authorities, and environmental permitting.

9.3.2 Pumped Refrigerant Cooling Systems

The pumped refrigerant approach to transporting the heat discussed above is not new; it is just now being implemented on a much larger scale than the thermosiphon from years ago. It utilizes a pure refrigerant in that it does not require the use of lubricants, such as mineral oil or POE oil, as does direct expansion systems. This eliminates the risk of fire hazard and eliminates the potential negative impact on various fire detection systems.

The theory of operation of the system is quite simple. There is a pump in the system to ensure the fluid continues to circulate even at light loads. The fluid begins its cycle as a liquid and as it absorbs heat in the heat absorption coils located very near the source of the heat, it will flash to a gas. Only a percentage (typically 50% to 85% at full heat load) will flash to a gas and return to the pumping unit. In the pumping unit, it is condensed back to a liquid either in a plate heat exchanger connected to the building chilled-water system or, in some cases, the heat exchanger is the evaporator for a chiller system. This would be a refrigerant-to-refrigerant heat exchanger. Once condensed back to 100% liquid, the pump recirculates the fluid back to the heat absorption coils.

The pumped refrigerant is pumped at a temperature above the room dew-point temperature to ensure no condensation occurs either on the piping or at the heat absorption coils at the load. If a leak were to occur, the fluid leaks as a gas, not a liquid.

Heat exchanger design is important for refrigerants and, if optimized, can result in significant energy-efficiency benefits. First, a two-phase fluid such as R-134a or R-744 does not change temperature as it absorbs the heat as long as it never reaches a 100% gaseous state. This means the entering fluid temperature can be higher than for a fluid that is not changing phase and still achieve the same net temperature of the device being cooled. Second, when utilizing a two-phase fluid process, the volume of fluid required can be greatly reduced versus a single-phase process (see Section 9.6 for more detail). This means the size of the tubes in the heat exchanger, sometimes referred to as *microchannel coils*, can be much smaller than with water for the same pressure drop. A two-fold improvement in cooling capability is typically attained. Third, since the heat exchanger coils are more efficient in transferring the heat, they allow for much lower air-side pressure drop—a two-fold reduction is very common.

9.3.3 Dielectric-Based Liquid Cooling Systems

Dielectric fluids may also take advantage of phase-change phenomena to enhance heat transfer. Examples of dielectric-based liquid cooling include immersion cooling, jet impingement cooling, and evaporative spray cooling. Immersion cooling was originally implemented in old supercomputer systems. Jet impingement cooling still appears to be in the research stage, and there are no reported commercially available systems that are cooled via jet impingement cooling. Evaporatively spray-cooled systems are commercially available today in some systems.

Evaporatively spray-cooled systems typically consist of cooling system software and an associated controller, a rack-based heat exchanger, a rack-based supply and return manifold, a reservoir, and a pump (Cader et al. 2005, 2007). Energy-efficient evaporative cooling is accomplished in a variety of ways.

Phase Change at the Liquid-Cooled Cold Plate and at the Heat Exchanger. The efficiency of this part of the process stems from the fact that large amounts of heat can be rejected at the heat exchanger with very low fluid inventory and within a very narrow temperature difference.

Low Coolant Inventory. At the system level, evaporatively cooled systems use significantly lower fluid inventory than immersion-cooled systems, and less than other single-phase systems.

Rejection to Chilled or Condenser Water. Systems have been designed such that the saturated two-phase mixture that exits the liquid-cooled cold plate is at 113°F (45°C); this leaves a 27°F (15°C) temperature difference to reject the waste heat to the condenser water at 86°F (30°C). By rejecting the waste heat directly to condenser water, data centers that have implemented evaporative cooling can either downsize their chiller plants, or can use the freed up chiller capacity for other parts of their building complexes. Figure 9.3 illustrates such an opportunity.

Reduced Airflow Rate Requirement. Evaporatively cooled systems typically reject 50% of a server's heat (from the microprocessors) directly to chilled water or condenser water.

Adjustable Coolant Flow Rate. This control at the rack level can facilitate more efficient part-load operation than trying to reduce airflow rates at part load with air-cooled systems.

9.4   LIQUID-COOLED ENCLOSURES

As mentioned in Section 9.1, there are two designations for liquid-cooled racks—open-cooling architectures and closed-cooling architectures. An open architecture is a room-cooling solution that effectively manages all of the air in a room, while a closed system manages the air within an enclosed cabinet. Both architectures operate on the same basic principal. Hot air is passed through an air-to-water or air-

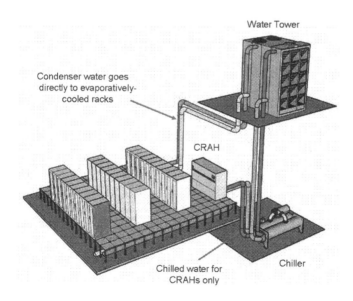

Figure 9.3   Diagram of waste heat rejection bypassing mechanical cooling.

to-refrigerant heat exchanger. Heat is transferred to the fluids and then sent to condensers, chillers, and, ultimately, out of the building. With the heat removed, the cooled air is then circulated back to the IT equipment air intakes at the front of the enclosure. Diagrams showing the various options for connecting liquid-cooled enclosures to a facility chilled-water or process-cooling-water system are provided in other ASHRAE publications (ASHRAE 2005d, 2006b) and are not repeated here.

9.4.1 Open-Cooling Architecture

Open-cooling architecture is available in three basic configurations, all of which use an open or open mesh front enclosure door. The configuration is usually independent of the fluid used to transport the heat. The configurations are as follows:

1. The hot air exiting the servers internal to the rack passes through a heat exchanger that removes the heat from the air prior to it entering the open aisle. This configuration in some cases eliminates the need for a hot-aisle/cold-aisle arrangement, as the heat can be fully neutralized before the air enters the aisle. Some systems are designed to utilize the server fans to transport the air through the heat exchanger, while others have their own fan mechanisms.

2. A heat exchanger/fan unit collects the heat from the air in the rear of the enclosures, cools it, and then transmits it into the cold aisle. By proper placement of the units, the cold aisle is effectively pressurized, preventing hot air from migrating into the cold aisle.

3. The hot aisle is contained via mechanical barriers, including a false ceiling. This prevents the hot air from escaping to the cold aisles and allows for a much higher hot-aisle temperature, thus increasing the potential capacity of the heat exchange units. Heat exchanger/fan units collect this hot air, cool it, and then transmit it into the cold aisle.

9.4.2 Closed-Cooling Architecture

Closed-cooling architecture is available in two basic configurations and, most commonly, is totally enclosed. In normal operation, the air internal to the enclosure is kept separate from the external ambient. The exception to this rule is during emergency operation caused by loss of cooling capacity. In this event, the various configurations provide some mechanism (opening doors or dampers) to allow the room air to enter the cabinet and provide extra time for orderly shutdown of the equipment internal to the cabinet. Two basic air circulation paths are typically maintained in the enclosures—horizontal (front to rear) or vertical (top to bottom). Both use multiple fans in either sidecar units (horizontal) or rear door mount (vertical) to move air through components. Heat exchangers are located in the sidecar units or in the bottom of the equipment enclosure. For water-cooled units, connections to building chilled-water supply and return are required. An alternative common approach is to connect the water-cooled units to a coolant distribution unit (CDU), which, in turn, is connected to the building chilled-water system. The purpose of the CDU is to (1) provide isolation between clustered groups minimizing the impact of a leak, (2) minimize the filtration requirement of the building chilled water, and (3) control the fluid temperature to eliminate the potential for condensation on the pipes, coils, and enclosure. Refrigerant systems also require inlet and outlet connections. Depending on the existing piping, number of units to be cooled, and expected heat load, connections to water-cooled cabinets can be provided directly from water supply and return pipes, with individual connections to each cabinet from underfloor pipe loops. Cabinets with refrigerant systems will have pipe runs from the enclosures back to a CDU, which, in turn, are connected to chilled-water systems.

Installation of liquid-cooled enclosures requires careful planning to minimize TCO. Additional floor space is needed to accommodate the sidecars (typically 12 in. wide) or the extra depth for the vertical systems (up to 52 in.). With floor space allocated, two strategies can be used for enclosure deployment:

1.   Designate a specific area within an existing facility for high-density installation.
2.   For new construction, and a few existing sites, consider a complete water-cooled solution.

Most data centers can benefit from assigning a certain amount of floor space for high-density/high-heat load components. This concentrates the hardware and services in one area, improving component accessibility, manageability and service. It also localizes infrastructure (piping, cable, power) required to support the loads, minimizing the impact on remaining floor space.

The second alternative, a complete liquid-cooled site, may at first seem impractical. However, for existing facilities that have no more thermal capacity, or are out of floor space with no room for expansion, this solution may be the only alternative. Overall facility floor space can be reduced. This solution can also make sense for remote installations in nontraditional spaces. More systems are being installed out of the data center—factory floor, warehouse, campus, closets, and other remote locations—and many of these spaces may not have the cooling infrastructure required to support network and server components. These nontraditional installations can often be best handled with the closed cooling architecture approach since they can be configured to be a "data center in a box."

Liquid-cooled cabinets provide a close-coupled system that places the liquid heat removal medium as close to the heat sources as practical, which can be designed to be quite efficient. Bypassing an air-cooling loop at the facility level saves on fan energy, and may increase the discharge temperature of the heat rejected to the environment, allowing for increased use of economizer cycles.

9.5   TCO ASPECTS OF LIQUID COOLING

Total cost of ownership (TCO) and return on investment (ROI) are important considerations in the overall analysis of liquid cooling systems. It may be more expensive to install liquid cooled systems than traditional passive or active air-cooled systems, but the potential energy savings and related costs savings may make these installations economically feasible. The increased heat removal capabilities of these systems can also mean that additional cost savings can be realized for future higher density and higher heat load installations, since facilities with liquid-cooled enclosures or chip-cooling technologies can typically support these future products with less impact on existing infrastructure.

When analyzing TCO and ROI for liquid cooling, many factors should be considered. These factors include the following:

- Hardware costs
- Warranty costs
- Coolant costs
- Space climate control
- Cost of electricity
- Floor space
- Heat density planned
- Flexibility required for future changes
- Cabling
- Lighting
- Maintenance costs
- Building costs

These factors are discussed in more detail below.

Hardware Costs. The cost of the datacom equipment and the cost of housing the equipment will be affected by the decision to pursue liquid cooling. This cost should be factored into a TCO analysis.

Warranty Costs. The cost to insure a liquid cooling system may be different than the corresponding cost of insuring an air-cooled system. This difference should be factored into the TCO analysis.

Coolant Costs. Some coolants cost more than others. The cost of providing the coolant, and the cost of coolant inventory in case of leakage, should be factored into the analysis.

Space Climate Control. Liquid-cooled systems at the rack level typically require significantly less room air conditioning, since heat is rejected to a liquid rather than to air circulating in the space. The capital and operating costs of liquid cooling thus need to be compared to the reduction in capital and operating costs of space-cooling systems. The liquid-cooling approaches can typically be optimized to perform only sensible cooling. This effort will reduce the size of the building chilled-water plant since the chiller plant is normally sized to accommodate the total gross capacity of the various CRAC units. If the liquid cooling is 100% sensible, there is no need to add humidity to the room caused by excess latent heat removal. Reducing the chiller plant size will also reduce the size of the emergency backup generators and corresponding switchgear.

Liquid cooling also allows for a reduction in the energy consumed by air-moving devices. As one gets closer to the actual source of the heat, fan energy should be reduced due to shorter duct runs and associated frictional losses.

Cost of Electricity. To fully understand the energy component of TCO, a complete energy analysis needs to be made to look at the relative costs of a given liquid-cooled system to alternative air-cooled systems and/or other liquid-cooled options.

Floor Space. A significant advantage of liquid cooling is that greater installed component density can typically be achieved. The sum of the enclosure footprint can be reduced by up to 70%, with concurrent savings in terms of reduced aisle and support floor space.

Heat Density Planned. Understanding and planning for future load density changes does not mean installing excess capacity at early stages, but rather allowing for the easy implementation of increased capacity at a later date when needed. This eliminates the installation and operation of excess capacity prior to its requirement.

Flexibility. Planning for and building in flexibility for future load changes can lead to significant capital cost and energy savings in both the early stages and the latter stages.

Cabling. Fewer equipment enclosures with higher component installation density per enclosure will typically reduce cable quantities (connectivity and power) installed in liquid-cooled facilities.

Lighting. Less floor space will typically also mean reduced lighting, with its associated capital and operating costs.

Maintenance Costs. If we reduce or eliminate the need for air recirculation, the impact of dust in the data center is correspondingly reduced. Liquid-cooling solutions generally do not require air filters that need to be maintained. Depending on the fluid used, fluid filters may or may not be required.

Overall Building Costs. If a facility designer were to consider all the points listed above, a liquid-cooling infrastructure may make it practical to build a smaller facility and reduce overall construction costs. The reader should perform a rigorous analysis to determine a site and project-specific comparison.

When trying to determine TCO for various options, it is important to note that line item comparisons will vary from site to site. The analysis can be affected by the following:

- Geographical location—north vs. south, warm vs. cold, humid vs. dry
- Data center location—upper vs. lower floor, south facing vs. north facing, windows vs. solid walls

- Planned heat density
- Flexibility required for future changes
- Facility construction—raised vs. solid floor, number and location of CRAC units, chilled water availability
- Energy costs—$/KWH

As the demand for faster processing and higher capacity storage continues to increase, with ever growing heat loads, the need for liquid-cooled systems will likely also increase. To optimize TCO and ROI for the owner, a full analysis of the capital and operating costs of liquid-cooling options must be performed.

9.6   A SAMPLE COMPARISON: PUMPING POWER

The properties of air, water and refrigerants are substantially different and, as a result, it is difficult to make an "apples-to-apples" comparison of these three fluids. A designer, however, must make a comparison between real-world fluids to determine which to use for a specific application. The intent of this section is not to indicate that one fluid is "better" than another, but simply to show how such a comparison might be made. It is up to the reader to input their own pressure drops, etc., for a specific application to determine the best fluid.

One of the most significant differences between air, water, and a typical refrigerant (R-134a in this example) is in their respective density at atmospheric conditions. Referring to Table 9.4, the density of air at standard conditions is 0.076 lb/ft^2 (1.217 kg/m^3), while the density of water is 62.4 lb/ft^2 (998 kg/m^3), and the density of R-134a, the refrigerant used in this example, is 76.6 lbs/ft^3 (1225 kg/m^3).

Another very important property is the specific heat of the fluid. The specific heat of air is 0.245 Btu/lb·°F (1026 J/kg·K), the specific heat of water is 1.004 Btu/lb·°F (4287 J/kg·K), and the specific heat of R-134a is 0.337 Btu/lb·°F (1403J/ kg·K). For this example, however, we have assumed that the heat exchanger is at a temperature that will allow the refrigerant to flash and change phase. In this case, the important parameter for the evaporating refrigerant is no longer the specific heat, but, rather, the heat of vaporization, which, in a relative sense, is quite high (93 Btu/lb [217,000 J/kg]). Even after multiplying the specific heat of air and water by reasonable expected temperature rises (22°F [12.2K] for air; 12°F [6.7K] for water), the volumetric heat transfer capacity of the phase-change liquid is still quite large compared to either air (17,400:1) or water (9.5:1).

From an energy transport perspective, the power required to pump a given volume of liquid water or refrigerant is much higher than for a given volume of air due to the higher density of the liquids. The comparison in Table 9.4 uses a representative fan pressure of 2.5 in. w.g. (0.623 kPa) for air, and pump pressures of 50 ft (15.25 m) for liquid, corresponding to about 150 kPa for water and 187 kPa for R-134a. In this example, the pressure needed to pump a given volume of liquid is greater than air by a ratio of between 240:1 to 300:1.

By dividing the volumetric heat capacity of air, water, and refrigerant by the respective amount of power needed to pump the fluids, an estimate of the coefficient

Table 9.4 Sample Comparison of Transport COP for Air, Water, and Phase-Change Refrigerant Heat Rejection

	Air	Water	R-134a
Density	0.076 lb/ft^3	62.3 lb/ft^3	75.3 lb/ft^3
Heat capacity	0.245 Btu/lb·°F	1.00 Btu/lb·°F	92.8 Btu/lb·°F (heat or vapor)
		3346 ratio	N/A
Volumetric heat capacity	0.01862 Btu/ft^3·°F	62.3 Btu/ft^3·°F	6988 Btu/ft^3
Typical heat rise	22°F	12°F	Phase change
Volumetric heat-transfer content	0.4096 Btu/ft^3	748 Btu/ft^3	6988 Btu/ft^3
		1825 ratio to air	17,059 ratio to air
Time period conversion	60 min/h	60 min/h	60 min/h
Flow rate	1.00 ft^3/min	1.00 ft^3/min	1.00 ft^3/min
Heat transfer per unit time	24.58 Btu/h	44,858 Btu/h	419,295 Btu/h
Typical HVAC system pressure drop	2.5 in. w.g.	50.0 ft w.g.	60.4 ft w.g.
Typical HVAC system pressure drop	13.01 lb/ft^2	3121 lb/ft^2	3772 lb/ft^2
		240 ratio	290 ratio to air
Required horsepower per 1 ft^3/min of flow rate	0.000393 hp/ft^3/min	0.0944 hp/ft^3/min	0.1141 hp/ft^3/min
Heat transfer	62,488 Btu/hp·h	475,198 Btu/hp·h	3,675,659 Btu/hp·h
Horsepower conversion	2545 Btu/hp·h	2545 Btu/hp·h	2545 Btu/hp·h
COP (energy removed/input prime mover energy input)	24.5 Btu/Btu	186.7 Btu/Btu	1444 Btu/Btu
Ratio (liquid:air)		7.60 ratio to air	58.8 ratio to air
Ratio (phase-change refrigerant:water)			7.74 ratio to water

of performance (COP) for transporting a given amount of heat is obtained. In this case, the COP refers to the ratio of the amount of heat transferred to the total electrical energy input to the pump (expressed as equivalent heat). With the values used for this example, the COP of air transport is about 25, the COP of water transport is about 187, and the COP of refrigerant transport is 1444. In this case, therefore, air is the least efficient means of heat transport, and refrigerant is the most efficient. It is important to note that Table 9.4 contains representative values, which are quite dependent on project-specific temperature rises and respective fan and pumping pressures. Every project will have its own design criteria, but as Table 9.4 shows, the differences in transport energy can be dramatic.

9.7   ENERGY-EFFICIENCY RECOMMENDATIONS/BEST PRACTICES

Facilities and systems to support liquid cooling will be required to address the demands of higher density datacom equipment, increased overall heat load, less real estate, and improved performance. The purpose of this chapter has been to provide an introduction to the technical and economic considerations that are integral to an examination of liquid-cooling systems in order to provide the reader with the tools necessary to make an informed decision. If the decision is made to utilize liquid cooling for some or all of a facility's cooling needs, the following points should be considered:

- Determine whether the cooling architecture will be open or closed.
- Choose a cooling fluid that is right for the project. In practice there will be many possible solutions; comparing key coolant properties, such as those shown in Table 9.2, will be helpful in making this decision.
- If a water-based liquid-cooling solution is chosen, the following parameters are all important:
    - Minimize pumping power.
    - Consider designs that maximize chiller efficiency and economizer hours of operation.
    - Provide for an efficient heat rejection design
- If a pumped refrigerant liquid-cooling solution is chosen, the following parameters are all important:
    - Choose the right fluid for the project.
    - Minimize pumping power.
    - Provide for an efficient heat exchanger design.
- If a dielectric liquid cooling solution is chosen, the following parameters are all important:
    - Choose the right fluid for the project—latent heat of vaporization is important.

- Minimize pumping power—including use of variable-speed pump controls.
- Provide for an efficient heat exchanger design.
- Consider heat rejection from the dielectric loop directly to the condenser water, saving on chilled-water costs and infrastructure.
- For most liquid-cooling solutions, the use of a CDU can allow for the liquid-cooling loop temperature to be set above the room dew-point temperature, eliminating condensation.
- Recognize that datacom equipment loads will change over the next 10 to 15 years. Develop a liquid-cooling strategy that can adjust to these changes.
- Calculate the TCO, and make a decision based on all the parameters relating to TCO.

10

Total Cost of Ownership

10.1   INTRODUCTION

Total cost of ownership (TCO) is a methodology that the datacom facility owner can use to make an economics-based decision between two or more options, or a comparison between their datacom facility and (1) another facility or (2) an industry benchmark. The keys to TCO are contained in the words that make up this acronym. Of the three words, *cost* is the main word. What is the cost, and how much must the owner pay for the data center? The costs must be balanced with the benefits the owner obtains. The word *total* implies that all aspects of the decision are being considered in the analysis, and all potential costs are being included; this includes future costs, such as operating costs and downtime, in addition to initial capital costs. The first step in *ownership* is obtaining the resource or the cost to buy or build it. Once it is owned, the datacom facility must be operated and maintained. All costs associated with the ownership of the datacom facility and its contents need to be considered.

Industry Focus on TCO

The datacom industry is increasingly scrutinizing TCO as corporations ask information technology (IT) and facilities departments to do more with less each year. Two factors make the need for in-depth TCO analysis important. First is the rising cost of energy. With electricity and fuel costs on the increase, the operational costs of operating a datacom facility are on the rise. Second, the increasing density available in the current generation of servers allows for a much greater increase in computational or service capability than even a few years ago. This good news brings along a challenge for the facility designer and owner to balance the opportunity for more capacity with installing the infrastructure to support the datacom equipment. The choices are extensive, and the analytical methods to make the right choice come down to an economic comparison between multiple options. When datacom equipment had a low power density, the equipment could go many more places without a significant infrastructure impact. The infrastructure impact today, however, can be a major part of choosing how much equipment, as well as even which equipment, to procure.

Benefits of the Analysis

The need for detailed TCO analysis is greater than ever in the datacom space. Large data centers can use vast amounts of power. As an example, consider a data center with 10,000 servers, with each server averaging 400 W over the course of a year. If we assume the electricity costs $0.10/kWh (near the national average), the annual energy bill to operate *just the servers* would be roughly $3.5 million. Minimizing this cost is important to the business and/or mission success.

The other important aspect of a TCO analysis is the benefit it provides the end user in providing a rigorous basis for comparing options or where to focus on reducing costs. Unfortunately, data center design and management has been regarded as an art rather than a science, with too much hearsay and speculation, based on often-repeated incorrect facts. A detailed TCO analysis has the opportunity to dispel many of the incorrect beliefs and permit the owner to make the right decisions and best populate their data center with the required performance at the lowest total cost.

10.2 TCO METHODOLOGY

10.2.1 Background

The methodology used to determine TCO is essentially the appropriate application of engineering economics. For a detailed overview, the reader is encouraged to first review the *2007 ASHRAE Handbook—Applications* (ASHRAE 2007b, 2007c, 20007d). The *Handbook* covers the detailed methodology, and this will not be repeated here in the same level of depth, although the concepts will be considered in their application to datacom facility and equipment decisions.

As mentioned previously, the TCO methodology needs to consider all appropriate costs. Generally these will include capital or first costs, as well as on going or recurring costs. The line item costs need to be summed to determine the TCO. The complicated part is dealing with the costs of recurring and future items (monthly energy bill, replacement of a component in several years, etc.) and calculating the present value of all these costs.

10.2.2 Time Value of Money (The Discount Rate)

One of the first steps in the TCO analysis is to determine the appropriate time value of money. This is often called the discount rate. This factor takes into consideration the fact that money on hand today has a greater value than the same sum of money will have at some future date.

The discount rate is set by the owner to determine their "time value of money" based on a number of factors. The owner's finance department is typically the best place to find out what discount rate should be applied. The discount rate represents the minimum return a company expects from the investments it makes, and is sometimes called the *hurdle rate*. Consider three different firms, each with a need to expand their datacom capabilities. The first firm is a small start-up, with very little cash on hand. The second is a services firm with a large amount of cash reserves. The third also has reasonable cash reserves, but there is a shortage on the market for their main product and, if they built a new factory, they could obtain a very high rate of ROI. Each firm has different expectations for the ROIs they make. The topic is complicated, and assistance and information from the owner's finance representative is generally the best source of information on the appropriate value to use.

In its simplest form, the discount rate could be applied as follows. Assume there is a maintenance contract to change filters in an air-side economizer and the cost is $1000.

$1000 dollars today is worth $1000.

If the contract is paid at the end of the year with a discount rate of $i = 6\%$, the present value of the contract is $943.

$$P = F\frac{1}{(1+i)^n}$$

where

P = present value
F = future value
i = discount rate
n = number of periods

If the contract is paid monthly ($1000/12 = $83.33/month$) with a discount rate of 0.5%/month, the present value of the contract is $968.

$$P = A\left[\frac{(1+i)^n - 1}{i(1+i)^n}\right]$$

where

A = equal payment value

Because this is a payment, the one time outlay of $1000 at the end of the year is preferable, as it has the least cost (lowest present value). This analysis is a very simple example of the concepts, and the reader is encouraged to look deeper into the references for more details (ASHRAE 2007b, 2007c, 2007d; ASTM 2006).

Taking the concepts further, it is obvious why there is a need to use a time value of money. Consider the implementation of the air-side economizer. The hardware will cost a certain amount more to purchase today. However, there is an expected savings each year in the energy bill for the power required to cool the data center. It is incorrect to simply add all of the annual savings together, as the value of the savings in the fifth year would not hold the same value (in today's dollars) as the first years' savings, even though the sum each year may be the same. This is a straightforward outcome from the discount factor discussion earlier in this section.

The other important concept is at what point in time to make the comparison. Convention and simplicity dictate that the options be compared using the present value of all costs. The alternate would be to do the analysis and the comparison at some future date in time. This method is as equally sound technically, but adds complication because one needs to determine what future time to use, as well as the need to convert all costs to these future values.

10.2.3 Life Time Questions

The present value of the options needs to be calculated in today's money. The strength in the TCO calculation is being able to fairly and accurately compare multiple options, including those that may span different time frames. For example, if there was an option that included a first cost and then an annual expense thereafter, and another with a lower first cost and annual expense but included a major maintenance item (equipment replacement) every three years, then it would be appropriate to do an evaluation that spanned a couple of the major costs in the future years to fairly capture those costs.

Life of the Facility and Infrastructure vs. Life of the Datacom Equipment

One challenge in datacom facility design, as well as TCO analysis of that design, is the fact that facilities are typically built with an expected life of 15 years or more. The building, power delivery, and cooling design are typically based upon current loads and space needs, with an assumed growth pattern. The refresh rate for the datacom equipment served, however, is closer to 3 years. A significant question is, thus, the proper time frame for TCO analysis.

Server technology generally follows Moore's Law, which states that the number of transistors in an integrated circuit will double roughly every 18 months. While Dr. Moore was not speaking directly about servers, the silicon inside the servers is following Moore's Law and is the primary driver of the compute capability increases in the data center. A liberal application of Moore's Law would indicate that over a 15-year life of a datacom facility, the servers being installed in year 15 would have more than 1000 times the compute capacity of the first year's servers. The challenge of designing a facility to support a thousand-fold increase is what makes the task so difficult. Fortunately the power required to drive the servers has risen at a rate orders of magnitude less than Moore's law, but the increase between the power draw of a typical server of 15 years ago versus today is very real.

10.2.4 Simplified TCO Modeling Methodology

The following represents a process to follow for the execution of a detailed TCO analysis. It assumes that the reader is already well versed in the engineering economic analyses, as detailed in the *2007 ASHRAE Handbook—Applications* (ASHRAE 2007b, 2007c, 2007d).

TCO Process

1. Identify the question to be answered or choice to be made.
2. Determine the appropriate discount rate.
3. Determine the time frame of the problem.
4. List and consider all direct impacts and associated costs.
5. Review potential second-order effects; list any that are "maybes."
6. List and calculate all first-cost items.
7. List all future or recurring costs.
8. Calculate the present value of all future expenditures or costs.
9. For each option, sum all costs to determine the TCO.
10. Compare options; make choice based on lowest TCO.

While each analysis will have its own unique challenges, maintaining a consistent methodology will help to streamline this process and provide the intended value from the TCO analysis.

What to Include: Just Enough Complexity to Answer the Question

One challenge in any TCO analysis is to include the right cost components. This brings into question the validity of the use of the word *total* in TCO. Perhaps it is more appropriate to consider *total* as describing a smaller subset of the facilities overall cost, depending on what question is being asked—or that the method will determine the difference in TCO between two options, but may not always yield the absolute value. Again, consider the example of an air-side economizer being analyzed to provide greater cooling efficiency and lower TCO. Option A is a standard system, with no air-side economizer. Option B incorporates all of option A but also includes the additional controls, ducting, and dampers, as well as filters and air-treatment needed to successfully implement the technology. The costs to include in this example would certainly include the capital costs of both options A and B, highlighting the added cost of the economizer. In addition, the costs of operating both systems (fans, chillers, pumps) needs to be analyzed, and these costs brought back to a present value, with (hopefully) option B indicating a lower annual operating cost and lower contribution to TCO. The analysis would also have to include the added cost of increased filter replacement and the manpower needed for the additional maintenance of the economizer system.

If a true "total" cost of ownership of the data center and infrastructure were to be calculated, the analysis would need to include the cost of the servers, the raised floor, the power to run the servers, the cost of the security system, the monthly fee for connectivity to the Internet, salaries for the operators, etc.; however, many of these have nothing to do with the choice between options A and B and which is better. Their addition in the analysis would only serve to complicate the review and add time and extra detail not needed in the decision-making process. The TCO methodology has been employed in a manner that efficiently calculates the difference in TCO between the two options, normalized to present value. One alternate way of looking at it is that the output of the analysis is the TCO of the fan/ventilation portion of the cooling system.

10.2.5 ROI as an Alternate Methodology

Using the discount rate for your company allows calculating the TCO of an investment that recognizes the time value of money. The results can be compared in today's dollars to select the lowest cost alternative. Another approach is to calculate the ROI for each alternative and compare them to your company's minimum acceptable rate of return. The ROI is the discount rate at which the net positive cash flows over time (savings or incremental profits), resulting from an investment equal to the initial investment. If the ROIs for all of the alternatives exceed the company's minimum acceptable return, then the alternative with the highest ROI is likely the best choice.

With ROI, a time frame must again be chosen. Many cost or energy reduction opportunities will pay for themselves eventually, but most owners are not willing to wait 5, 10, or 15 years for a payback, both due to more attractive short-term investment opportunities and because the risk of change (see next section) adds too much uncertainty to long-range investments.

10.2.6 Uncertainty and Risk

There are several risks with TCO calculations. These are (1) not comparing like situations, (2) missing a factor in the analysis, (3) predicting future values, and (4) awareness of potential disruptive technologies.

The Common Base Line

It is important to have a common base line when doing TCO analyses that is based on the actual applied usage. Some of the criteria to consider making common between all of the scenarios being evaluated are as follows:

1. The desired supply and return conditions of temperature and humidity (unless these are specific variables in the parametric analysis)

2. The control dead bands and other operating characteristics
3. The ambient "bin data" being used for ambient conditions
4. The load factor of the room (most rooms have a minimum of 20% redundancy even on the worst design-day scenario)

Missing Factor Risk

The missing factor risk can only be avoided through a detailed consideration of all of the impacts of a change or alternate option, including seeking input from others associated with the datacom facility. A recommended approach is to enlist a cross-functional team with members from groups impacted by the decision, i.e., IT engineering and operations, facilities, finance, etc. The multiple different groups involved also make this challenge worse. The facilities department owns and operates the building and infrastructure. The IT department is responsible for server and software selection, installation, and operation. The site procurement group is often responsible for the electricity bill, rate negotiations, and leases. Finally, if the question includes siting or location issues, a corporate real estate office is often involved. All too often these groups are unaware of each other's issues and costs. Bridging the gaps between the various disciplines is mandatory to ensure that all factors are accounted for.

Predicting Future Values

A company's finance department will typically provide the discount rate to use in the TCO analysis. Uncertainty related to future energy costs can be addressed by analyzing scenarios at two or more assumed energy cost levels. This approach will demonstrate the sensitivity of the decision to energy costs. If the results are the same regardless of which case is used, the decision is simple. If, on the other hand, option X is better with the best-case energy cost assumption, while Option Y is the lower TCO with the worst-case assumption, a judgment call will be required on the likelihood of the future costs, related to these limits. In addition, unanticipated market factors can come into play. One example would be a shortage of raw material, such as copper or steel.

Potential Disruptive Technologies

Finally, potential disruptive technologies could come into play to alter the results of a well laid out TCO analysis. These types of changes could be anything from server manufacturers implementing liquid cooling to the microprocessor in the server, to virtualization significantly impacting server utilization, resulting in much higher-power density, but better efficiency performance.

10.3   TCO ANALYSIS AS APPLIED TO
ENERGY COSTS AND ENERGY-EFFICIENCY METRICS

10.3.1 Calculation of Annual Energy Costs

To determine the TCO of various energy options, the first calculation that must be made is the expected annual energy consumption. This calculation requires detailed technical expertise on component and system interactions. Typically a model of the mechanical systems is created and meshed with weather data files and datacom equipment load variations to determine the annual energy usage. The most comprehensive software to determine this annual energy usage is the US Department of Energy's (USDOE) EnergyPlus program (USDOE 2007). More simplified modeling techniques using binned weather data can also be used. The USDOE maintains a list of software tools on its Web site—currently, over 300 are listed (USDOE 2007).

Once the energy consumption is known, it can be multiplied by the energy rate schedule for specific time bins (on-peak/off-peak, time of year, etc.) to determine the annual facility cost. If the energy costs are different from year to year, it may be possible to apply escalation factors in the analysis program. If, however, there are step functions to the energy load side (such as from a staged build out), separate model runs may be required for different years to obtain energy cost information in sufficient detail to add to the TCO analysis.

10.3.2 Energy-Efficiency Metrics (On an Annualized Basis)

Once annual energy consumption is known, this information can be reduced to more simple metrics to allow for comparison with industry standards.

For instance, an annualized data center efficiency (DCE) or power usage effectiveness (PUE) (see Chapter 1) can be calculated to compare the various options. The annualized DCE or PUE should include the following:

- The use of air- and water-side economizers, which vary with environmental conditions
- Variations in IT power draw, due to fluctuating workload
- Time period over which the power measurements are made
- Electrical rate structure
- VFDs
- External climate and environmental conditions
- Building envelope and ancillary loads, such as lighting
- Control networks (SCADA, BMS. etc.)
- Impact of use of space (office, warehouse, IT, food service, etc.)

These metrics are more clearly defined and easier to obtain where the power feed is completely dedicated to the datacom facility. However, in multipurpose or multi-floor buildings where one or more of the floors is used for general administration and office space, the electrical and cooling load supporting these areas needs to be considered and eliminated from the total building load in order to establish an accurate DCE or PUE rating.

10.4   TCO ANALYSIS AS APPLIED TO NON-ENERGY COSTS

Since the emphasis of this publication is on energy and energy efficiency, detailed analysis of non-energy costs is not included. For IT equipment, most manufacturers have guidelines that will guide the analysis.

References already stated provide a good overview of all the factors that are needed for a complete TCO analysis. For quick reference, however, some of the additional factors that would come into play in a complete facility and IT equipment TCO analysis include the following:

Capital costs
    Land
    Design costs
    Power delivery
    Cooling system
    IT equipment
    Safety and security
    Network and connectivity

Operational
    Rental/lease
    Power
    UPS
    Power distribution
    Cooling

Other considerations
    Taxes
    Depreciation
    Incentives
    Funding source
    Software
    Disposal fees
    Lost revenue from downtime
    Customer fees/payments for hosting

An online calculator that can help to make certain TCO analyses is available at the time of this publication (Koomey 2007b).

10.5   SUMMARY

TCO calculations are an important tool in making financial decisions in datacom facilities. The most important considerations in conducting a TCO analysis, particularly for decisions related to energy consumption, include the following:

- Generate a list of valid TCO options, based on research, discussion with experts, and interactions with all disciplines and departments. Determining the best option early in the design process is one of the best ways to minimize TCO.
- Find out the discount rate and time frame for analysis.
- Run through the TCO methodology listed in Section 10.2.4, paying particular attention to the "complexity" discussion to make sure that the items included and level of detail match the questions being asked.
- For energy comparisons, a software model that incorporates the physical attributes of the facility, the mechanical systems, the datacom load variations, the local weather data, and the energy costs and escalation factors are needed to provide a quality analysis.

11

Emerging Technologies and Future Research

11.1   INTRODUCTION

The intent of this chapter is to provide an overview of some of the newer technologies that are being examined that may have a substantial impact on the energy efficiency and/or TCO of datacom facilities.

11.2   DC POWER

A conventional datacom facility utilizes an AC power distribution infrastructure. However, the electronic components within the datacom equipment typically require DC power to operate.

Consequently, a conversion from AC power to DC power is required, and this is typically accomplished within the individual power supplies to each piece of datacom equipment. However, this is not the only power conversion that occurs within the power distribution system. Because of the high power requirements of datacom facilities, utility power is often provided at medium voltage (12 kV or more); it is then transformed to 480 V for distribution within the facility.

The power distribution infrastructure will often include UPS units to help isolate the datacom equipment from power outages and power quality disturbances. The UPS units use DC battery strings to store energy and, consequently, they incorporate a rectifier (to convert AC to DC) and an inverter (to convert DC to AC). The 480 V AC power is ultimately transformed to 120 or 208 V before being distributed to the datacom equipment racks.

As illustrated in Figure 11.1, there could be six or more stages of power conversion before the utility power is delivered to the datacom electronics. Each conversion is a potential source of power loss and wasted energy, therefore making it important to try to reduce the number of times the power needs to be converted and to improve the efficiency of each power conversion process.

There can be a notable difference in the efficiency of UPS unit, and datacom facility owners and engineers should balance first cost against lifetime cost when

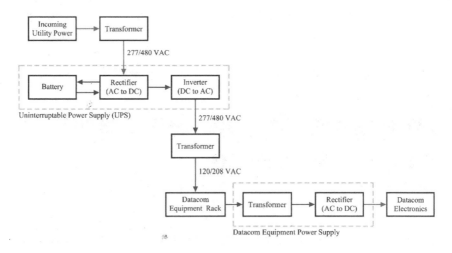

Figure 11.1 Schematic representation of a typical AC power distribution system for a datacom facility.

selecting these devices. Similarly, there is a wide range in the performance and efficiencies of datacom equipment power supplies.

One method to reduce the number of conversion processes is to utilize a DC power distribution scheme in lieu of an AC power distribution scheme. 48V DC power distribution systems are fairly common within telecom facilities, but the power cabling requirements can become burdensome for high-density datacom equipment. Although not common practice, a recent study (LBNL 2007d) highlighted the benefits of adopting a 380 V DC power distribution system for a datacom facility as a means to improve overall energy efficiency when compared to traditional 480 V AC power distribution systems. Figure 11.2 shows a potential efficiency gain of about 7% for a DC power distribution system when compared to two AC power distribution systems.

11.3   FUEL CELLS, DISTRIBUTED GENERATION, AND CHP

A traditional data center typically uses the power from the utility company as the primary source of power under normal conditions. In the event of a utility outage, a backup power source (typically an on-site generator) provides power on a temporary or emergency basis until the utility power is restored and is considered stable.

Distributed generation (DG) is a term that describes having a locally generated primary power source operating on a continuous basis. DG technologies include solar photovoltaic, wind turbines, engine generators, turbine generators and fuel cells. Some of these technologies have a suitable waste product (heat) that

System Efficiency	UPS Efficiency	Transformer Efficiency	PS Efficiency	System Efficiency
AC System A: Measured Efficiency	90%	98%	90%	79%
AC System B: Measured Efficiency	90%	98%	90%	79%
DC System A: Measured Efficiency	94%	100%	92%	87%

Energy Consumption	Compute Load (kWh)	Input Load (kWh)	Efficiency Gain
AC System A: Measured Consumption	23.3	26.0	
AC System B: Measured Consumption	23.3	25.9	
DC System A: Measured Consumption	22.7	24.1	
% Energy Consumption Improvement vs. AC System A			7.3%
% Energy Consumption Improvement vs. AC System B			7.0%

Figure 11.2 Comparison of AC and DC power distribution system efficiencies (LBNL 2007d).

can be used as the input energy for the cooling as a part of a combined heat and power (CHP) system.

Many studies exist describing the energy savings of a DG/CHP system as the primary source, with savings thought to be in the neighborhood of 35% over a traditional utility power scenario. Reliability can also be increased by having the utility power act in a secondary source capacity. Another consideration is the cleaner emission of a DG/CHP system that uses fuel cells.

A fuel cell is an electrochemical device that combines hydrogen and oxygen to produce electricity with a waste product of water and heat. The conversion of the fuel to energy takes place via an electrochemical process, and can be two to three times more efficient than fuel burning.

11.4   FUTURE RESEARCH

Research to develop a more energy-efficient data center is an ongoing endeavor, and it is difficult to know in advance what paths might yield the best improvements in efficiency. Research in the following areas, however, is likely to yield more energy efficient datacom facilities.

- Continued benchmarking, with very specific characterizations of facilities and breakouts of energy use, so that we can better compare various designs, and continue to quantify the opportunities for energy-efficiency improvements and trends that are occurring over time.

- Development of a number of "standard datacom facility" profiles that can be used for energy-efficiency analysis. This analysis would include the examination of annual energy consumption in a number of locations and with a number of designs, to better understand the trade-offs offered by alternatives such as different supply air temperatures, different types of economizers, and liquid vs. air cooling.
- Continued improvements in the efficiency of mechanical cooling equipment.
- Continued improvements in the efficiency of electrical distribution equipment.
- Continued improvements in the efficiency of datacom equipment.

11.5   CONCLUSIONS

As this book has hopefully demonstrated, energy consumption in the datacom environment is multifaceted. There is no silver bullet that will start to reduce datacom facility energy consumption, and the successful "architect" of an energy-efficient datacom facility will be one who can mobilize talent in many disciplines and integrate many of the ideas contained in this book (as well as new ideas that evolve with time) into a successful strategy for energy efficiency.

A summary of the major recommendations presented in this book can be found in Table 1.1.

Bibliography and References

BIBLIOGRAPHY

Air Force. 1978. Facility design and planning engineering weather data, AFM 88-29, July.

ANSI. 1997. *ANSI T1.304-1997, Ambient Temperature and Humidity Requirements for Network Equipment in Controlled Environments*. New York: American National Standards Institute. www.ansi.org.

ANSI. 2003. ANSI draft (2003), Engineering Requirements for a Universal Telecommunication Framework (UTF), Standards Committee T1 Telecommunications, Working Group T1E1.8, Project 41.

APC. 2003. Alternative power generation technologies for data centers and network rooms. American Power Conversion, #64, Revision 1.

Azar, K. 2002. Advanced cooling concepts and their challenges. *Proceedings of the 8th International Workshop on Thermal Investigation of ICs and Systems, October 1–4, 2002*. www.qats.com.

Bash, C., C. Patel, and R. Sharma. 2004. Efficient thermal management of data centers—Immediate and long-term research needs. *ASHRAE Transactions*.

Beaty, D., and T. Davidson. 2003. New guideline for data center cooling. *ASHRAE Journal*. 45(12):28–34.

Beaty, D. 2004. Liquid cooling—Friend or foe. ASHRAE Symposium June 2004.

Beaty, D. 2004. Specifying underfloor plenums. *HPAC Engineering*, November.

Beaty, D. 2005. Commissioning combined with CFD for datacom performance, adaptability, and reliability. *ASHRAE Transactions*.

Beaty, D. 2005. Commissioning raised-floor plenums. *HPAC Engineering*, January.

Beaty, D. 2005. Cooling data centers with raised-floor plenums. *HPAC Engineering*, September.

Beaty, D., and R. Schmidt. 2004. Back to the future—Liquid cooling: Data Center Considerations. *ASHRAE Journal* 46(12):42–48.

Belady, C. 2001. Cooling and power considerations for semiconductors into the next century, [invited]. *Proceedings of the International Symposium on Low Power Electronics and Design*, August.

Belady, C., and D. Beaty. Data centers: Roadmap for datacom cooling. *ASHRAE Journal* 47(12):52–55.

Chauhan, N., D. Beaty, and D. Dyer. 2005. High density cooling of data centers and telecom facilities—Part 1. ASHRAE Symposium February 2005.

Chauhan, N., D. Beaty, and D. Dyer. 2005. High density cooling of data centers and telecom facilities—Part 2. ASHRAE Symposium February 2005.

Chu, R.C. 2003. The challenges of electronic cooling: Past, current and future. *Proceedings of IMECE: International Mechanical Engineering Exposition and Congress, November 15–21, 2003, Washington DC.*

Cohen, J.E., and C. Small. 1998. Hypsographic demography: The distribution of human population by altitude. *Proceedings of the National Academy of Sciences of the United States of America* 95(24):14009–14, November 24.

Conner M., and L. Hannauer. 1988. Computer center design. *ASHRAE Journal*, April, pp. 20–27.

Davidson, T., and D. Beaty. 2005. Data centers: Datacom airflow patterns. *ASHRAE Journal*, April.

EIA. 1992. EIA-310, revision D, Sept. 1, 1992: Racks, panels and associated equipment.

Ellis, M. 2002. Status of fuel cell systems for combined heat and power applications in buildings. ASHRAE Transactions 108(1):1032–44.

Engle, D. 2005. Power sufficiency with "chilling" efficiency. *Distributed Energy Journal*, July/August.

Eto, J.H., and C. Meyer. 1988. The HVAC costs of fresh air ventilation. *ASHRAE Journal*, September, pp. 31–35.

ETSI. 1994. ETSI 300 019-1-0 (1994-05) Equipment Engineering (EE); Environmental conditions and environmental tests for telecommunications equipment Part 1-0: Classification of environmental conditions, www.etsi.org.

ETSI. 1999. ETSI EN 300 019-2-3 V2.1.2 (1999-09) Equipment Engineering (EE), Environmental conditions and environmental tests for telecommunications equipment; Part 2-3: Specification of environmental tests; Stationary use at weather protected locations, www.etsi.org.

Felver, T.G., M. Scofield, and K. Dunnavant. 2001. Cooling California's computer centers. *HPAC Heat, Piping, and Air Conditioning Engineering* 73(3)59–63.

Garday, D., and D. Costello. 2006. Air-cooled high-performance data centers: Case studies and best methods, www.intel.com/IT.

Garner, S.D. 1996. Heat pipes for electronics cooling applications. *Electronics Cooling Magazine*.

Guggari, S., D. Agonafer, C. Belady, and L. Stahl. A hybrid methodology for the optimization of data center room layout. *Proceedings of InterPACK '03, July 6–11.*

Hydeman, M., R. Seidl, and C. Shalley. 2005. Data centers: Staying on-line: Data center commissioning. *ASHRAE Journal* 47(4):60–65.

ITRS. 2003. *International Technology Roadmap For Semiconductors,* 2003 ed.

Kang, S., R. Schmidt, K.M. Kelkar, A. Radmehr, and S.V. Patankar. 2001. A methodology for the design of perforated tiles in raised floor data centers using computational flow analysis. *IEEE Transactions on Components and Packaging Technologies.*

Karki, K.C., A. Radmehr, and S.V. Patankar. 2003. Use of computational fluid dynamics for calculating flow rates through perforated tiles in raised-floor data centers. *International Journal of HVAC&R Research* 9(2).

Kiff, P. A fresh approach to cooling network equipment. Published under "Technical Papers" at www.flovent.com.

Kurkjian, C., and J. Glass. 2005. Air conditioning design for data centers—Accomodating current loads and planning for the future. *ASHRAE Transactions* 111(2):715–24.

Lawson, S.H. 1988. Computer facility keeps cool with ice storage. *Heating, Piping, Air Conditioning* 60(8):35–38, 43, 44.

Liebert Engineering White Paper. 2003. Utilizing economizer effectively in the data center, www.liebert.com

Mills, M. 1999. *The Internet Begins with Coal: A Preliminary Exploration of the Impact of the Internet on Electricity Consumption.* Greening Earth Society, Arlington, VA.

Mitchell, R.L. 2003. Moving toward meltdown. *Computer World Magazine.*

Mitchell-Jackson, J., J. Koomey, and M. Blazek. 2001. National and regional implications of Internet data center growth, Draft copy, December 11, 2001.

Nakao, M., H. Hayama, and M. Nishioka. 1991. Which cooling air supply system in better for a high heat density room: Underfloor or overhead? *Proceedings of the Thirteenth International Telecommunications Energy Conference (INTELEC '91), November.*

Noh, H.-K., K.S. Song, and S.K. Chun. 1998. The cooling characteristics on the air supply and return flow systems in the telecommunication cabinet room. *Proceedings of the Twentieth International Telecommunications Energy Conference (INTELEC '98), October.*

Patel C., R. Sharma, C. Bash, and A. Beitelmal. 2002. Thermal considerations in cooling large scale compute density data centers. *Proceedings of the ITherm Conference, June.*

Patel, C., Bash, C. Belady, L. Stahl, and D. Sullivan. 2001. Computational fluid dynamics modeling of high density data centers to assure systems inlet air specifications. Accepted for publication in the *Proceedings of InterPACK 2001 Conference, ASME,* July.

Patterson, M., R. Steinbrecher, and S. Montgomery. 2005. Data centers: Comparing data center & computer thermal design. *ASHRAE Journal* 47(4):38-42.

PNL. 1990. *Architect's and Engineer's Guide to Energy Conservation in Existing Buildings*, Volume 2, Chapter 1. DOE/PL/ 01830 P-H4. Pacific Northwest Laboratories.

Prisco, J. 2006. Contamination sources and prevention in data processing environments. *ASHRAE Transactions*.

Quivey, B., and A.M. Bailey. 1999. Cooling of large computer rooms—Design and construction of ASCI 10 TeraOps. InterPack 1999.

Rodgers, T. 2005. An owner's perspective on commissioning of critical facilities. *ASHRAE Transactions* 11(2):618–26.

Schmidt, R., C. Belady, A. Classen, T. Davidson, M. Herrlin, S. Novotny, and R. Perry. 2004. Evolution of data center environmental guidelines. *ASHRAE Transactions*.

Schmidt, R., and E. Cruz. 2002. Raised floor computer data center: Effect on rack inlet temperatures of chilled air exiting both the hot and cold aisles. ITherm conference, June.

Schmidt, R. and E. Cruz. 2002. Raised floor computer data center: Effect on rack inlet temperatures when high powered racks are situated amongst lower powered racks. IMECE conference, November.

Schmidt, R. and E. Cruz. 2003. Clusters of high powered racks within a raised floor computer data center: Effect of perforated tile flow distribution on rack inlet air temperatures. IMECE Conference, November. To be published.

Schmidt, R. and Cruz, E., 2003. Raised Floor Computer Data Center: Effect on rack inlet temperatures when adjacent racks are removed. Interpack Conference, July 2003.

Schmidt, R. and B. Notohardjono. 2002. High-end server low-temperature cooling. *IBM J. Res. Develop.* 46(6).

Schmidt, R. 2001. Effect of data center characteristics on data processing equipment inlet temperatures, advances in electronic packaging. *Proceedings of IPACK '01, The Pacific Rim/ASME International Electronic Packaging Technical Conference and Exhibition*, Vol. 2, Paper IPACK2001-15870, July.

Schmidt, R. 2001. Water cooling of electronics. Cooling Electronics in the Next Decade Sponsored by Cooling Zone, August.

Schmidt, R. 2004. Thermal profile of a high density data center-methodology to thermally characterize a data center. *ASHRAE Transactions*.

Schmidt, R., M. Iyengar, and R. Chu. 2005. Data centers: Meeting data center temperature requirements. *ASHRAE Journal* 47(4):44-48. Atlanta.

Schmidt, R., K.C. Karki, K.M. Kelkar, A. Radmehr, and S.V. Patankar. 2001. Measurements and predictions of the flow distribution through perforated tiles in raised-floor data Centers. *Proceedings of InterPack'01, July*.

Silverling, A.M., and K.J. Kressler. 1995. Ice storage system assures data center cooling. *HPAC Heating, Piping, Air Conditioning* 67(4):35–39.

Sorell, V., Y. Abougabal, K. Khankari, V. Gandi, and A. Watve. 2006. An analysis of the effects of ceiling height on air distribution in data centers. *ASHRAE Transactions*.

Sorell, V., S. Escalante, and J. Yang. 2005. A comparison of underfloor and above-floor air delivery systems in a data center environment using CFD modeling. *ASHRAE Transactions*.

Ståhl, Lennart. 2004. Cooling of high density rooms: Today and in the future. *ASHRAE Transactions*.

Telcordia. 2001. GR-3028-CORE, Thermal management in telecommunications central offices. *Telcordia Technologies Generic Requirements*, Issue 1, December. Piscataway, NJ: Telcordia Technologies, Inc.

Telcordia. 2002. GR-63-CORE, Network equipment—Building system (NEBS) requirements: Physical protection. *Telcordia Technologies GenericRequirements*, Issue 2, April. Piscataway, NJ: Telcordia Technologies, Inc.

Vukovic, A. 2004. Communication network power efficiency—Assessment, limitations and directions. *Electronics Cooling Magazine*, August.

Vukovic, A. 2005. Data centers: Network power density challenges. *ASHRAE Journal*. 47(4):55–59.

Yamamoto, M., and T. Abe. 1994. The new energy-saving way achieved by changing computer culture (saving energy by changing the computer room environment). *IEEE Transactions on Power Systems*, Vol. 9, August.

REFERENCES

80PLUS. 2006. http://www.80plus.org/. October 20, 2006.

AIA. 2003. Best practices—Building commissioning and maintenance. BP 19-01-01. December. Washington, DC: American Institute of Architects.

APC. 2003. Calculating total cooling requirements for data centers. White Paper #25. West Kingston, RI: American Power Conversion.

ASHRAE. 1991. *ASHRAE Terminology of Heating, Ventilation, Air Conditioning, and Refrigeration*. Atlanta: American Society of Heating, Refrigerating and Air-Conditioning Engineers, Inc.

ASHRAE. 1992. *ANSI/ASHRAEStandard 52.1-1992, Gravimetric and Dust-Spot Procedures for Testing Air-Cleaning Devices Used in general Ventilation for Removing Particulate Matter*. Atlanta: American Society of Heating, Refrigerating and Air-Conditioning Engineers, Inc.

ASHRAE. 1996. *ASHRAE Guideline 1-1996, The HVAC Commissioning Process*. Atlanta: American Society of Heating, Refrigerating and Air-Conditioning Engineers, Inc.

ASHRAE. 1999. *ANSI/ASHRAE Standard 52.2-1999, Method of Testing General Ventilation Air-Cleaning Devices for Removal Efficiency by Particle Size*. Atlanta: American Society of Heating, Refrigerating and Air-Conditioning Engineers, Inc.

ASHRAE. 2001. *ANSI/ASHRAE Standard 127-2001, Method of Testing for Rating Computer and Data Processing Room Unitary Air Conditioners.* Atlanta: American Society of Heating, Refrigerating and Air-Conditioning Engineers, Inc.

ASHRAE. 2003. *2003 ASHRAE Handbook—HVAC Applications.* Chapter 17, "Data Processing and Electronic Office Areas." Atlanta: American Society of Heating, Refrigerating and Air-Conditioning Engineers, Inc.

ASHRAE. 2004a. *ANSI/ASHRAE/IESNA Standard 90.1-2004, Energy Standard for Buildings Except Low-Rise Residential Buildings.* Atlanta: American Society of Heating, Refrigerating and Air-Conditioning Engineers, Inc.

ASHRAE. 2004b. *2004 ASHRAE Handbook—HVAC Systems and Equipment.* Chapter 12, "Hydronic Heating and Cooling System Design." Atlanta: American Society of Heating, Refrigerating and Air-Conditioning Engineers, Inc.

ASHRAE. 2004c. *2004 ASHRAE Handbook—HVAC Systems and Equipment.* Chapter 18, "Fans." Atlanta: American Society of Heating, Refrigerating and Air-Conditioning Engineers, Inc.

ASHRAE. 2004d. *2004 ASHRAE Handbook—HVAC Systems and Equipment.* Chapter 19, "Evaporative Air Cooling Equipment." Atlanta: American Society of Heating, Refrigerating and Air-Conditioning Engineers, Inc.

ASHRAE. 2004e. *2004 ASHRAE Handbook—HVAC Systems and Equipment.* Chapter 20, "Humidifiers." Atlanta: American Society of Heating, Refrigerating and Air-Conditioning Engineers, Inc.

ASHRAE. 2004f. *2004 ASHRAE Handbook—HVAC Systems and Equipment.* Chapter 36, "Cooling Towers." Atlanta: American Society of Heating, Refrigerating and Air-Conditioning Engineers, Inc.

ASHRAE. 2004g. *2004 ASHRAE Handbook—HVAC Systems and Equipment.* Chapter 43, "Heat Exchangers." Atlanta: American Society of Heating, Refrigerating and Air-Conditioning Engineers, Inc.

ASHRAE. 2004i. *ANSI/ASHRAE Standard 34-2004, Designation and Safety Classification of Refrigerants.* Atlanta: American Society of Heating, Refrigerating and Air-Conditioning Engineers, Inc.

ASHRAE. 2004j. *ANSI/ASHRAE Standard 55-2004, Thermal Environmental Conditions for Human Occupancy.* Atlanta: American Society of Heating, Refrigerating and Air-Conditioning Engineers, Inc.

ASHRAE. 2004k. *ANSI/ASHRAE Standard 62.1-2004, Ventilation for Acceptable Indoor Air Quality.* Atlanta: American Society of Heating, Refrigerating and Air-Conditioning Engineers, Inc.

ASHRAE. 2004l. *2004 ASHRAE Handbook—HVAC Systems and Equipment.* Chapter 38, "Liquid Chilling Systems." Atlanta: American Society of Heating, Refrigerating and Air-Conditioning Engineers, Inc.

ASHRAE. 2005a. *2005 ASHRAE Handbook—Fundamentals.* Chapter 15, "Fundamentals of Control." Atlanta: American Society of Heating, Refrigerating and Air-Conditioning Engineers, Inc.

ASHRAE. 2005b. *2005 ASHRAE Handbook—Fundamentals*. Chapters 23, 24, 25, and 30. Atlanta: American Society of Heating, Refrigerating and Air-Conditioning Engineers, Inc. (All referenced chapters contain information related to building envelope design.)

ASHRAE. 2005c. *Datacom Equipment Power Trends and Cooling Applications*. Atlanta: American Society of Heating, Refrigerating and Air-Conditioning Engineers, Inc.

ASHRAE. 2005d. *Design Considerations for Datacom Equipment Centers*. Atlanta: American Society of Heating, Refrigerating and Air-Conditioning Engineers, Inc.

ASHRAE. 2005e. *ASHRAE Guideline 0-2005, The Commissioning Process*. Atlanta: American Society of Heating, Refrigerating and Air-Conditioning Engineers, Inc.

ASHRAE. 2005f. *Weather Data Viewer*, Version 3.0. Atlanta: American Society of Heating, Refrigerating and Air-Conditioning Engineers, Inc.

ASHRAE. 2006a. Addendum h to *ANSI/ASHRAE/IESNA Standard 90.1-2004, Energy Standard for Buildings Except Low-Rise Residential Buildings*. Atlanta: American Society of Heating, Refrigerating and Air-Conditioning Engineers, Inc.

ASHRAE. 2006b. *2006 ASHRAE Handbook—Refrigeration*. Chapter 1, "Absorption Cooling, Heating, and Refrigeration Equipment." Atlanta: American Society of Heating, Refrigerating and Air-Conditioning Engineers, Inc.

ASHRAE. 2006c. *Liquid Cooling Guidelines for Datacom Equipment Centers*. Atlanta: American Society of Heating, Refrigerating and Air-Conditioning Engineers, Inc.

ASTM. 2006. *E917-05, Standard Practice for Measuring Life-Cycle Costs of Buildings and Building Systems*. Philadelphia, PA: American Society of Testing and Materials.

ASHRAE. 2007a. *2007 ASHRAE Handbook—Applications*. Chapter 17, "Data Processing and Electronic Office Areas." Atlanta: American Society of Heating, Refrigerating and Air-Conditioning Engineers, Inc.

ASHRAE. 2007b. *2007 ASHRAE Handbook—Applications*. Chapter 35, "Energy Use and Management." Atlanta: American Society of Heating, Refrigerating and Air-Conditioning Engineers, Inc.

ASHRAE. 2007c. *2007 ASHRAE Handbook—Applications*. Chapter 36, "Owning and Operating Costs." Atlanta: American Society of Heating, Refrigerating and Air-Conditioning Engineers, Inc.

ASHRAE. 2007d. *2007 ASHRAE Handbook—Applications*. Chapter 38, "Operation and Maintenance Management." Atlanta: American Society of Heating, Refrigerating and Air-Conditioning Engineers, Inc.

ASHRAE. 2007e. *2007 ASHRAE Handbook—Applications*. Chapter 40, "Building Energy Monitoring." Atlanta: American Society of Heating, Refrigerating and Air-Conditioning Engineers, Inc.

ASHRAE. 2007f. *2007 ASHRAE Handbook—Applications*. Chapter 41, "Supervisory Control Strategies and Optimization." Atlanta: American Society of Heating, Refrigerating and Air-Conditioning Engineers, Inc.

ASHRAE. 2007g. *2007 ASHRAE Handbook—Applications*. Chapter 51, "Evaporative Cooling." Atlanta: American Society of Heating, Refrigerating and Air-Conditioning Engineers, Inc.

ASHRAE. 2007h. *2007 ASHRAE Handbook—Applications*. Chapter 59, "Codes and Standards." Atlanta: American Society of Heating, Refrigerating and Air-Conditioning Engineers, Inc.

ASHRAE. 2007i. *ANSI/ASHRAE Standard 127-2007, Method of Testing for Rating Computer and Data Processing Room Unitary Air Conditioners*. Atlanta: American Society of Heating, Refrigerating and Air-Conditioning Engineers, Inc.

ASHRAE. 2007j. *Structural and Vibration Guidelines for Datacom Equipment Centers*. Atlanta: American Society of Heating, Refrigerating and Air-Conditioning Engineers, Inc.

ASHRAE. 2009. *Thermal Guidelines for Data Processing Environments, Second Edition*. Atlanta: American Society of Heating, Refrigerating and Air-Conditioning Engineers, Inc.

ASTM. 2006. *E917-05, Standard Practice for Measuring Life-Cycle Costs of Buildings and Building Systems*. Philadelphia, PA: American Society of Testing and Materials.

Barry, M. 2004. Design issues in regulated and unregulated intermediate bus converters. *Applied Power Electronics Conference (APEC) 2004* 3:1389–94.

Belady, C. 2007. In the data center, power and cooling costs more than the IT equipment it supports. *Electronics Cooling* 13(1).

Cader, T., and K. Regimbal. 2005. Energy smart data center. InterPACK05 Panel on High Density Microprocessor Cooling. *Proceedings of InterPACK05, San Francisco, CA.*

Cader, T., K. Regimbal, L. Westra, and R. Mooney. 2005. Airflow management in a liquid-cooled data center. *ASHRAE Transactions* 112(2)220–30.

Cader, T., L. Westra, A. Marquez, H. McAllister, and K. Regimbal. 2007. Performance of a rack of liquid-cooled servers. *ASHRAE Transactions* 113(1)101–14.

Calm, J. and D. Didion. 1997. Trade-offs in refrigerant selections: Past, present and future. *Proceedings of ASHRAE/ NIST Refrigerants for the 21st Century Conference.*

CDA. 2003. Premium-efficiency motors and transformers. CD-ROM. New York: Copper Development Association.

CEMEP. 2002. UPS efficiency saving europe's power. European Committee of Manufacturers of Electrical Machines and Power Electronics, United Kingdom. http://www.upsci.com/pdf/UPS-efficiency.pdf.

Cinato, P., c. Bianco, L. Licciardi, F. Pizzuti, M. Antonetti, and M. Grossoni. 1998. An innovative approach to the environmental system design for TLC rooms in Telecom Italia. INTELEC.

Conner, M., and P. Hannauer. 1988. Computer center design. *ASHRAE Journal* 30(4):20–27.

DeDad, J.A. 1997. *Practical Guide to Power Distribution for Information Technology Equipment*. Overland Park, Kansas: Intertec Publishing Corporation.

Dou, S., W. Wu, A. Pratt, and P. Kumar. 2006. DC Transformer with Line Frequency Ripple Cancellation. *Proceedings of the International Power Electronics and Motion Control Conference (IPEMC) 2006* 1:537–41.

EIA. 1992. *ANSI/EIA-310-D-1992, Cabinets, Racks, Panels, and Associated Equipment*. Arlington: Electronic Industries Alliance.

ENERGY STAR®. 2007. ENERGY STAR program requirements for computers.

Fortenbury, B., and J.G. Koomey. 2006. Assessment of the impacts of power factor correction in computer power supplies on commercial building line losses. California Energy Commission, Contract # 500-04-030, March 2006.

Graves, B., R. Wanex, and C. Mamane. 1985. High efficiency equipment effects on short circuit currents. *Proceedings of the Technical Association of the Pulp and Paper Industry, pp. 69–74.*

The Green Grid. 2007. Green grid metrics: Describing data center power efficiency. Technical Committee White Paper. http://www.thegreengrid.org/downloads/Green_Grid_Metrics_WP.pdf.

Henderson, H., et al. 2003. An hourly building simulation tool to evaluate hybrid desiccant system configuration options. *ASHRAE Transactions* 109(2):551–64.

Herrlin, M.K. 1996. Economic benefits of energy savings associated with: (1) Energy-efficient telecommunications equipment; and (2) Appropriate environmental control. *Proceedings of Intelec'96, Boston, MA, October 6–10.*

Herrlin, M.K. 1997. The pressurized telecommunications central office: IAQ and energy consumption. Healthy Buildings/IAQ'97, Washington DC, Sept. 27–Oct. 2.

Herrlin, M.K. 2005a. Rack cooling effectiveness in data centers and telecomcentral offices: The Rack Cooling Index (RCI). *ASHRAE Transactions* 111(2):725–31.

Herrlin, M.K. 2005b. Survey presented at the 7th HDDC meeting (www.cfroundtable.org) at the National Energy Research Scientific Computing Center (NERSC), Oakland CA, June 3, 2005.

Herrlin, M.K., and C. Belady. 2006. Gravity-assisted air mixing in data centers and how it affects the rack cooling effectiveness. ITherm 2006, San Diego, CA, May 30–June 2.

IEEE. 1998. Recommended practice for establishing transformer capability when supplying nonsinusoidal load currents. Institute of Electrical and Electronics Engineers, Inc.

Intel. 2006. Intel NetStructure® MPCBL0040 Single Board Computer Technical Product Specification. August.

Intel. 2007. Effect of input voltage on PSU efficiency. Lab measurements.

Kiff, P. 1995. A fresh approach to cooling network equipment. *British Telecommunications Engineering*, July, pp. 149–55.

Koomey, J.G., et al. November 2006. Server energy measurement protocol: Version 1.0. http://www.energystar.gov/ia/products/downloads/Finalserverenergyprotocol-v1.pdf. February 14, 2007.

Koomey, J.G. 2007a. Estimating total power consumption by servers in the U.S. and the world. http://enterprise.amd.com/Downloads/svrpwrusecompletefinal.pdf. February 20, 2007.

Koomey, J.G. 2007b. Online calculator for calculating cost of ownership for data centers. http://uptimeinstitute.org/content/view/21/55/.

LBNL. 2005 High performance buildings: Data centers server power supplies. Lawrence Berkeley National Laboratory. Interim report.

LBNL. 2007a. Benchmarking: data centers—Charts. http://hightech.lbl.gov/benchmarking-dc-charts.html. Lawrence Berkeley National Lab.

LBNL. 2007b. Data center economizer contamination and humidity study. Lawrence Berkeley National Lab.

LBNL. 2007c. *Self Benchmarking Guide for Data Center Energy Performance*, Version 1.0., http://hightech.lbl.gov/documents/DATA_CENTERS/Self_benchmarking_guide-2.pdf, Lawrence Berkeley National Lab.

LBNL. 2007d. DC Power for Improved Data Center Efficiency, Lawrence Berkeley National Lab, January. http://hightech.lbl.gov/documents/DATA_CENTERS/DCDemoFinalReportJan17-07.pdf.

Lee, F.C., M. Xu, S. Wang, and B. Lu. 2006. Design challenges for distributed power systems. *International Power Electronics and Motion Control Conference (IPEMC) 2006* 1:1–15.

Lentz, M.S. 1991. Adiabatic saturation and VAV: A prescription for economy and close environmental control. *ASHRAE Transactions*, pp. 477–85 (refer to discussion at the end of the paper).

Malone, C.G., and C. Belady. 2006,. Metrics to characterize data center & IT equipment energy use. *Proceedings of Digital Power Forum, Richardson, TX*.

Marquet, D., et al. 2005. New flexible powering architecture for integrated service operators. *Proceedings of IEEE Intelec Conference*, pp. 575–80.

Mohapatra, S.C. 2006. An overview of liquid coolants for electronics cooling. *Electronics Cooling* 12(2).

NEMA. 1996. *Guide for Determining Energy Efficiency for Distribution Transformers*. No. TP-1. National Electrical Manufacturers Association, Rosslyn, VA.

NEMI. NEMI Technology Roadmaps, 2000, 2002 and 2004 eds. National Electronics Manufacturing Initiative.

NFPA. 2005. *NFPA 70: National Electrical Code Handbook.* Quincy, MA: National Fire Protection Agency.

OSHA. 2007. Technical Manual Section III, Ch. 4, Table III:4-2 http://www.osha.gov/dts/osta/otm/otm_iii/otm_iii_4.html.

Pette, M.A. 1996. Quantifying energy savings from commissioning: Preliminary results from the Northwest. *Proceedings of the National Conference on Building Commissioning.*

Rasmussen, N. 2006. The advantages of row and rack-oriented cooling architectures for data centers. White paper #130. http://www.apcc.com/prod_docs/results.cfm?class=wp&allpapers=1.

Ren, Y., M. Xu, K. Yao, and F.C. Lee. 2003. Two-stage 48V power pod exploration for 64-bit microprocessor. *Applied Power Electronics Conference (APEC) 2003*, Vol. 1, pp. 426–31.

Schmidt, R.R., R.C. Chu, M. Ellsworth, M. Iyengar, D. Porter, V. Kamath, and B. Lehman. 2005. Maintaining datacom rack inlet air temperatures with water-cooled heat exchanger. *Proceedings of Pacific Rim ASME International Electronic Packaging Technical Conference and Exhibition (IPACK 2005), San Francisco, California, July 17–22.*

Sharma, R.K, C.E. Bash, and C.D. Patel. 2002. Dimensionless parameters for evaluation of thermal design and performance of large-scale data centers. *Proceedings of 8th ASME/AIAA Joint Thermophysics and Heat Transfer Conference, St. Louis, Missouri, June 24–26, 2002.*

Saidi, M.H., et al. 2007. Hybrid desiccant cooling systems. *ASHRAE Journal* 49(1):44–49.

SPEC. 2007. http://www.spec.org/specpower/. February 22, 2007. Standard Performance Evaluation Corporation.

Sullivan, R. 2007. The impact of Moore's Law on the total cost of computing and how inefficiencies in the data center increase these costs. *ASHRAE Transactions* 113(1):457–61.

Telcordia. 2001. Generic Requirements NEBS GR-3028-CORE, Thermal Management in Telecommunications Central Offices, Issue 1, December 2001, Telcordia Technologies, Inc., Piscataway, NJ.

Telcordia. 2002. Generic Requirements NEBS GR-63-CORE, NEBS Requirements: Physical Protection, Issue 2, 2002, Telcordia Technologies, Inc., Piscataway, NJ.

Ton, M., and B. Fortenbury. 2005. High performance buildings: Data centers uninterruptible power supplies (UPS). California Energy Commission's Public Interest Energy Research (PIER), December. http://hightech.lbl.gov/documents/UPS/Final_UPS_Report.pdf.

Ton, M., and B. Fortenbury. 2005. High performance buildings: Data centers server power supplies. Lawrence Berkeley National Laboratory. http://hightech.lbl.gov/documents/PS/Final_PS_Report.pdf.

Ton, M., and B. Fortenbury. 2006. High performance buildings: DC power for improved data center efficiency. Lawrence Berkeley National Laboratory, October. http://hightech.lbl.gov/documents/DATA_CENTERS/DCDemoFinalReportJan17-07.

US DoC. 1983. Guideline on Electrical Power for ADP Installations. United States Department of Commerce, National Bureau of Standards (Institute for Computer Sciences and Technology), Washington, DC.

USDOE. 2007. Building Energy Software Tools Directory. http://www.eere.energy.gov/buildings/tools_directory/.

USEPA. 2007. Report to Congress on Server and Data Center Energy Efficiency. Public Law 109-431. ENERGY STAR Program. August 2, 2007. U.S. Environmental Protection Agency.

VanGilder, J.W., and R.R. Schmidt. 2005. Airflow uniformity through perforated tiles in a raised-floor data center. *Proceedings of Pacific Rim ASME International Electronic Packaging Technical Conference and Exhibition (IPACK 2005), San Francisco, California. July 17–22, 2005.*

Vukovic, A. 2003. Power density challenges of next generation telecommunication networks. *Electronics Cooling* 9(1):34–40.

Vukovic, A. 2004. Communication network power efficiency—Assessment, limitations and directions. *Electronics Cooling* 10(3):18–24.

Vukovic, A. 2005. Network power-density challenges. *ASHRAE Journal* 47(4):55–59.

Weschler and Shields. 1998. Are indoor air pollutants threatening the reliability of your electronic equipment? *HPAC Magazine.* May.

Yester, R.J. 2006. New approach to high availability computer power system design. *Electrical Design, Construction, & Maintenance* 105(1):18–24.

Yong, Li, K. Sumathy, Y.J. Dai, J.H. Zhong, and R.Z. Wang. 2006. Experimental study on a hybrid desiccant dehumidification and air conditioning system. *Journal of Solar Energy Engineering* 128:77–82. February.

SELECTED WEB LINKS

80PLUS. 2006. http://www.80plus.org/. October 20, 2006.

Green Grid 2007. http://www.thegreengrid.org/.

Herrlin, M.K. 2005b. Survey presented at the 7th HDDC meeting (www.cfroundtable.org) at the National Energy Research Scientific Computing Center (NERSC), Oakland CA, June 3, 2005.

LBNL. 2007. http://hightech.lbl.gov/DCTraining/tools.html.

LBNL. 2007. http://hightech.lbl.gov/dc-benchmarking-results.html.

OSHA. 2007. Technical Manual Section III, Ch. 4, Table III:4-2 http://www.osha.gov/dts/osta/otm/otm_iii/otm_iii_4.html.

Rasmussen, N. 2004. Determining Implementing energy efficient data centers. White paper #114. http://www.apcc.com/prod_docs/results.cfm?class=wp&allpapers=1.

Rasmussen, N. 2005. Determining total cost of ownership for data center and network room infrastructure. White paper #6. http://www.apcc.com/prod_docs/results.cfm?class=wp&allpapers=1.

Rasmussen, N. 2006. The advantages of row and rack-oriented cooling architectures for data centers. White paper #130. http://www.apcc.com/prod_docs/results.cfm?class=wp&allpapers=1.

Standard Performance Evaluation Corporation (SPEC). 2007. http://www.spec.org/specpower/. February 22, 2007.

USDOE. 2007. EnergyPlus Energy Simulation Software. http://www.eere.energy.gov/buildings/energyplus/.

USEPA. 2007. Report to Congress on Server and Data Center Energy Efficiency. Public Law 109-431. U.S. Environmental Protection Agency. ENERGY STAR Program. http://www.energystar.gov/../EPA_Datacenter_Report_Congress_Final1.pdf.

Whole Building Design Guide. http://www.wbdg.org/design/lcca.php.

Appendix A

Glossary of Terms

A: current, amps (as used in Chapter 7).

A: equal payment value (as used in Chapter 10).

AC: alternating current.

adiabatic process: a thermodynamic process during which no heat is extracted from or added to the system.

air:

> *conditioned air:* air treated to control its temperature, relative humidity, purity, pressure, and movement.
>
> *exhaust air:* air extracted from a space, and expelled from the building without any recirculation.
>
> *return air:* air extracted from a space, and totally or partially returned to an air conditioner, furnace, or other heat source.
>
> *relief air:* air relieved from a space to the outside of a building, typically to maintain proper space pressurization during air-side economizer operation.
>
> *supply air:* air entering a space from an air-conditioning, heating, or ventilating apparatus; supply air is typically also conditioned air.
>
> *ventilation air:* air that is drawn in from the outside of the facility for the purposes of (1) diluting internal contaminants including CO_2, (2) maintaining facility pressurization, and/or (3) providing cooling during air-side economizer operation.

air-side economizer: see *economizer, air.*

availability: a percentage value representing the degree to which a system or component is operational and accessible when required for use.

BAS: building automation system—centralized building controls, typically for the purpose of monitoring and controlling environment, lighting, power, security, fire/life safety and elevators.

binned weather data: weather data that has frequency of incidence (often in units of hours per month or hours per year) occurring within discrete temperature ranges or bins.

BJT: bipolar junction transistor.

blanking panels: panels typically placed in unallocated portions of enclosed IT equipment racks to prevent internal recirculation of air from the rear to the front of the rack.

BMS: building management system (see *BAS*).

Btu: British thermal units; the amount of heat required to raise one pound of water one degree Fahrenheit; a common measure of the quantity of heat.

buck converter: step-down converter.

cabinet: frame for housing electronic equipment that is enclosed by doors and is standalone; this is generally found with high-end servers.

cabinet air: air (typically for the purposes of cooling) that passes through a cabinet housing datacom equipment.

CAF: conductive anodic failures.

central office (CO): telecom facility for mainly housing switching equipment. The physical, environmental, and electrical features differ significantly from those of traditional data center facilities.

CFD: computational fluid dynamics.

CHP: combined heat and power.

chilled-water system: an air- or process-conditioning system containing chiller(s), water pump(s), a water piping distribution system, chilled-water cooling coil(s), and associated controls. The refrigerant cycle is contained in a remotely located water chiller. The chiller cools the water, which is pumped through the piping system to the cooling coils.

chiller: a machine used to produce chilled water, typically for either air cooling and dehumidification, or liquid cooling. A chiller usually has a refrigerant cycle internal to it with two heat exchangers. The refrigerant evaporates in the evaporator, which is where the warm return water is cooled. The heat is then rejected in the condenser, either to a condenser water system (water cooled chiller), or directly to air (air-cooled chiller).

CMOS: complementary metal oxide silicon.

coefficient of performance (COP): (1) ratio of the rate of net heat output to the total energy input expressed in consistent units and under designated rating conditions and (2) ratio of the refrigerating capacity to the work absorbed by the compressor per unit time (ASHRAE 1991).

cold plate: typically a plate with cooling passages through which liquid flows to remove the heat from the electronic component to which it is attached.

commissioning: the process of ensuring that systems are designed, installed, functionally tested, and capable of being operated and maintained to perform in conformity with the design intent and begins with planning and includes design, construction, start-up, acceptance and training, and can be applied throughout the life of the building.

condenser: a heat exchanger in which vapor is liquefied by the rejection of heat to a heat sink. If the heat is rejected to air, the heat exchanger is called an air-cooled condenser. If the heat is rejected to water, it is called a water-cooled condenser.

conditioned air: see *air, conditioned.*

cooling coil:

> *chilled water:* a chilled-water cooling coil is an air-to-water cooling coil that utilizes chilled water (typically supplied by a chiller) to cool and/or dehumidify air.

> *evaporative:* an evaporative cooling coil is an air-to-refrigerant cooling coil, wherein a refrigerant is evaporated in the coil to cool and/or dehumidify air.

cooling tower: a heat-transfer device, often tower-like in shape, in which atmospheric air cools warm water, generally by direct contact (evaporation).

CRAC: computer room air conditioner; generally refers computer-room cooling units that utilize dedicated compressors and refrigerant cooling coils rather than chilled-water coils.

CRAH: computer room air handler; generally refers computer-room cooling units that utilize chilled-water coils for cooling rather than dedicated compressors.

DC: direct current.

DCE: data center efficiency; $\text{DCE} = \dfrac{\text{IT Equipment Power}}{\text{Total Facility Power}}$ (Green Grid 2007).

data center: a building or portion of a building whose primary function is to house a computer room and its support areas; data centers typically contain high-end servers and storage products with mission-critical functions.

datacom: a term that is used as an abbreviation for the data and communications industry.

dead band: the range of values within which a sensed variable can vary without initiating a change in the controlled process. A dead band is used in thermostats and humidistats to prevent excessive oscillation (called *hunting* in proportional control systems).

dehumidification: the process of removing moisture from the air in a space. It is often accomplished by cooling the air to a temperature below its dew point, at which time moisture condenses on a cooling coil, and is removed from the air stream via a drain.

DG: distributed generation.

dielectric fluid: a fluid that is a poor conductor of electricity.

direct digital control (DDC): a type of control where controlled and monitored analog or binary data (e.g., temperature, contact closures) are converted to digital format for manipulation and calculations by a digital computer or microprocessor, and then converted back to analog or binary form to control physical devices.

discount rate: the rate used for adjusting the total present economic value of a resource, projected over time, that takes into account the declining value of money.

diversity: a factor used to determine the load on a power or cooling system based on the actual operating output of the individual equipment, rather than the full load capacity of the equipment.

dry bulb: see *temperature, dry bulb.*

dry cooler: a heat rejection device that rejects heat to the environment without utilizing the evaporation of water. The fluid in a dry cooler does not change state. If it does, the term *air-cooled condenser* applies. (see also *wet cooler*).

dual enthalpy: refers to real-time measurement of both outdoor air and return air enthalpy, to ensure that the enthalpy of the outdoor air is lower than the enthalpy of the return air before an economizer cycle is activated.

DX (direct expansion) system: a system in which the cooling effect is obtained directly from the refrigerant. It typically incorporates a compressor, and in most cases the refrigerant undergoes a change of state in the system.

economizer, air: a duct and damper arrangement and automatic control system that together allow a cooling system to supply *outdoor air* to reduce or eliminate the need for mechanical cooling during mild or cold weather. Also called *air-side economizer.*

economizer, integrated: an economizer that allows for partial, but not complete, load reduction for mechanical cooling. Typically the air or water is precooled by the integrated economizer, and then is cooled to setpoint by the mechanical cooling equipment.

economizer, water: a system by which the supply air of a cooling system is cooled indirectly with water that is itself cooled by heat or mass transfer to the environment without the use of mechanical cooling. Also called *water-side economizer.*

efficiency: is "the ratio of the energy output to the energy input of a process or a machine" (ASHRAE 1991).

EGW: ethylene glycol/water.

electronically commutated motor (ECM): an EC motor is a DC motor with a shunt characteristic. The rotary motion of the motor is achieved by supplying the power via a switching device—the so-called *commutator.* On the EC motors, this commutation is performed using brushless electronic semiconductor modules.

EMI: electromagnetic interference.

energy conservation: more effective use of energy resources. Energy conservation seeks to reduce energy invested per unit of product output, service performed, or benefit received through waste reduction. Energy conservation and energy use reduction are not synonymous (ASHRAE 1991).

energy efficiency ratio (EER): the ratio of net equipment cooling capacity in Btu/h to total rate of electric input in watts under designated operating conditions. When consistent units are used, this ratio becomes equal to COP. (See also *coefficient of performance.*)

equipment: refers to, but not limited to, servers, storage products, workstations, personal computers, and transportable computers; may also be referred to as *electronic equipment* or *IT equipment.*

ESD: electrostatic discharge; increased risk of electronic equipment failure at low relative humidity levels.

exhaust air: see *air, exhaust.*

f: frequency.

F: future value.

fuel cell: an electrochemical device that combines hydrogen and oxygen to produce electricity with a waste product of water and heat.

GR: generic requirements documents from Telcordia (Bellcore); de-facto telecom standards.

h: abbreviation for hour.

HDF: hygroscopic dust failures; combination of hygroscopic dust on circuit boards and high relative humidity create conditions for current leakage and equipment failure.

heat:

> *total heat (enthalpy):* a thermodynamic quantity equal to the sum of the internal energy of a system plus the product of the pressure-volume work done on the system.

$$h = E + pv$$

> where h = enthalpy or total heat content, E = internal energy of the system, p = pressure, and v = volume.
>
> For the purposes of this document, h = sensible heat + latent heat.
>
> *sensible heat:* heat that causes a change in temperature.
>
> *latent heat:* change of enthalpy during a change of state.

heat exchanger: a device to transfer heat between two physically separated fluids.

heat pipe: also defined as a type of heat exchanger. Tubular closed chamber containing a fluid in which heating one end of the pipe causes the liquid to vaporize and transfer to the other end where it condenses and dissipates its heat. The liquid that forms flows back toward the hot end by gravity or by means of a capillary wick.

HEPA: high efficiency particulate air. These filters are designed to remove 99.97% or more of all airborne pollutants 0.3 microns or larger from the air that passes through the filter. There are different levels of cleanliness, and some HEPA filters are designed for even higher removal efficiencies and/or removal of smaller particles.

HOH (horizontal overhead): an air-distribution system is used by some long-distance carriers in North America. This system introduces the supply air horizontally above the cold aisles, and is generally utilized in raised-floor environments where the raised floor is used for cabling.

hot aisle/cold aisle: a common means of providing cooling to datacom rooms in which IT equipment is arranged in rows and cold supply air is supplied to the cold aisle, pulled through the inlets of the IT equipment, and exhausted to a hot aisle to minimize recirculation of the hot exhaust air with the cold supply air.

hot gas: pressurized gas leaving the compressor (discharge) prior to entering the condensing surface.

hp: horsepower.

humidity:

> *dew point temperature:* the temperature at which water vapor has reached the saturation point (100% relative humidity).
>
> *relative humidity (RH):* See *relative humidity (RH)*.

HVAC: heating, ventilation, and air conditioning.

hydronic: a term pertaining to water used for heating or cooling systems.

i: discount rate.

I: current.

infiltration: the flow of outdoor air into a building through cracks and other unintentional openings and through the normal use of exterior doors for entrance and egress. Also known as *air leakage into a building*.

infrastructure: the utilities and other physical support systems needed to operate a datacom facility. Included are electric distribution systems, cooling systems, water supply systems, sewage disposal systems, and roads.

IT: information technology.

kWh: kilowatt hour.

latent heat: see *heat, latent.*

lb: pound.

liquid cooling: conditioned liquid is supplied to the inlets of the rack/cabinet/server for thermal cooling of the heat rejected by the components of the electronic equipment within the rack. It is understood that within the rack, the transport of heat from the actual source component (e.g., CPU) within the rack itself can be either liquid or air based (or any other heat transfer mechanism), but the heat rejection media to the building cooling device outside of the rack is liquid.

mainframe: a high-performance computer made for high-volume, processor intensive computing. This term is used for the processor unit including main storage, execution circuitry and peripheral units, usually in a computer center, with extensive capabilities and resources to which other computers may be connected so they can share facilities.

MERV: minimum efficiency reporting value; a rating of the effectiveness of air filters in removing particle contaminants from the air. See ASHRAE (1999).

MOSFET: metal oxide silicon field effect transistor.

n: number of periods.

NEBS: a set of physical, environmental, and electrical de-facto telecom standards developed by Telcordia Technologies, Inc. (Bellcore).

NEMI: National Electronics Manufacturing Initiative.

OSHA: Occupational Safety and Health Administration.

P: present value.

Pascal (PA): A unit of pressure equal to one Newton per square meter. As a unit of sound pressure, one Pascal corresponds to a sound pressure level (SPL) of 94.

perforated floor tile: a tile as part of a raised floor system that is engineered to provide airflow from the cavity underneath the floor to the space. Tiles may be with or without volume dampers.

PDN: power delivery network.

PDU: power distribution unit.

performance specification: a specification in which performance requirements and required system outcomes are specified, but the detailed system design is not (see also *prescriptive specification*).

PFC: power factor correction.

PoL: point of load converter.

prescriptive specification: a specification in which detailed component and system design is specified rather than the desired outcome of the design (see also *performance specification*).

present value (or net present value): the current value of a future payment, or stream of payments, discounted at some appropriate compound interest, or discount, rate; also called *time value of money.*

PSU: power supply unit.

psychrometric chart: the psychrometric chart is a graph of the properties (temperature, relative humidity, etc.) of air. It is used to determine how these properties vary as the amount of moisture (water vapor) in the air changes.

PUE: power usage effectiveness; $PUE = \dfrac{\text{Total Facility Power}}{\text{IT Equipment Power}}$ (Green Grid 2007).

PWB: printed wire board.

R: resistance.

rack-mounted equipment: equipment that is to be mounted in an EIA (Electronic Industry Alliance) or similar cabinet; these systems are generally specified in EIA units, such as 1U, 2U, 3U, etc., where 1U = 1.75 in. (44 mm).

raised floor: a platform with removable panels where equipment is installed, with the intervening space between it and the main building floor used to house the interconnecting cables and at times is used as a means for supplying conditioned air to the information technology equipment and the room.

RCI: rack cooling index; measure of compliance with equipment intake temperature specifications such as ASHRAE and NEBS. RCI = 100% mean ideal conditions with no over- or under-temperatures. See Herrlin (2005a) for more detail.

refrigerants: in a refrigerating system, the medium of heat transfer which picks up heat by evaporating at a low temperature and pressure, and gives up heat on condensing at a higher temperature and pressure.

reliability: reliability is a percentage value representing the probability that a piece of equipment or system will be operable throughout its mission duration. Values of 99.9% (three 9s) and higher are common in data and communications equipment areas. For individual components, the reliability is often determined through testing. For assemblies and systems, reliability is often the result of a mathematical evaluation based on the reliability of individual components and any redundancy or diversity that may be employed.

relative humidity (RH): (1) ratio of the partial pressure or density of water vapor to the saturation pressure or density, respectively, at the same dry-bulb temperature and barometric pressure of the ambient air; and (2) ratio of the mole fraction of water vapor to the mole fraction of water vapor saturated at the same temperature and barometric pressure; at 100% relative humidity, the dry-bulb, wet-bulb, and dew-point temperatures are equal.

RHI: return heat index; $RHI = \dfrac{\text{Total heat extraction by the CRAC units}}{\text{Total enthalpy rise at the rack exhaust}}$; SHI + RHI = 1. See Sharma (2002) for more detail.

refrigerant condenser: a heat exchanger in which refrigerant vapor is liquefied by the rejection of heat to a heat sink. The heat sink medium can be either air cooled or liquid cooled.

refrigerating coefficient of performance (COP): ratio of the rate of heat removal to the rate of energy input in consistent units, for a complete refrigerating plant or some specific portion of the plant under designated operating conditions (ASHRAE 1991).

return air: see *air, return.*

sensible heat: see *heat, sensible.*

sensible heat ratio: the ratio of the sensible heat to the sensible plus latent heat to be removed from a conditioned space.

server: a computer that provides some service for other computers connected to it via a network. The most common example is a file server that has a local disk and services requests from remote clients to read and write files on that disk.

SHI: supply heat index; $\text{SHI} = \dfrac{\text{Enthalpy rise due to infiltration in cold aisle}}{\text{Total enthalpy rise at the rack exhaust}}$; SHI + RHI = 1. See Sharma (2002) for more detail.

supply air: see *air, supply.*

switchgear: the combination of electrical disconnects and/or circuit breakers used to isolate equipment.

telecom: abbreviation for *telecommunications.*

temperature:

> *dew point:* the temperature at which water vapor has reached the saturation point (100% relative humidity).

> *dry bulb:* the temperature of air indicated by a thermometer.

> *wet bulb:* the temperature indicated by a psychrometer when the bulb of one thermometer is covered with a water-saturated wick over which air is caused to flow at approximately 4.5 m/s (900 ft/min) to reach an equilibrium temperature of water evaporating into air, where the heat of vaporization is supplied by the sensible heat of the air.

thermal storage tank: a container used for the storage of thermal energy. Thermal storage systems are often used as a component of chilled-water systems.

thermostatic: of or relating to a thermostat, such as *thermostatic control.*

total cost of ownership (TCO): total cost of ownership (TCO) is a financial estimate (originating in the datacom industry) designed to assess direct and indirect costs related to the purchase of any capital investment. A TCO assessment ideally offers a final statement reflecting not only the cost of purchase but all aspects in the further use and maintenance of the equipment, device, or system considered.

TSS: thermal storage system.

turn down ratio: typically refers to the ratio of flows over which a control valve or other control device can reliably operate. Turn down ratio = maximum flow / minimum flow.

UPS: uninterruptible power supply.

VAV: variable air volume.

ventilation air: see *air, ventilation.*

VFD: variable frequency drive.

VOC: volatile organic compounds; may harm electronic components if not diluted with outdoor ventilation.

VOH: vertical overhead air distribution; dominant mode in telecom central offices.

VR: voltage regulator.

VSD: variable-speed drive.

VUF: vertical underfloor air distribution (raised floor); dominant mode in data centers.

water-side economizer: see *economizer, water.*

W: watts.

wet-bulb temperature: see *temperature, wet bulb.*

wet cooler: a heat rejection device that rejects heat to the environment by utilizing the evaporation of water. In same cases water is sprayed onto the coils to increase the heat transfer coefficient of the coil, in addition to the spray decreasing the ambient temperature through evaporation (see also *dry cooler*).

in. w.g.: pressure in inches of water gauge.

X_L: inductive reactance.

Z: impedance.

zone: a space or group of spaces within a building with heating and cooling requirements that are sufficiently similar so that desired conditions (e.g., temperature) can be maintained throughout using a single sensor (e.g., thermostat or temperature sensor).

Ω: impedance or resistance, ohms.

π: Pi (3.14159).

ρ: material specific resistance.

η: energy efficiency.

Appendix B

Commissioning, Maintenance, and Operations

INTRODUCTION

In a mission critical facility, housing sensitive datacom equipment, a loss of cooling can result in a severe loss of revenue, whether it results from a fault in the cooling plant upstream or in the secondary HVAC equipment installed locally in the data room. Loss of cooling usually results from mechanical breakdown or malfunctioning of building support systems. Electrical faults due to load mismatch, faulty wiring, or hazards are also common causes of failure and interruptions in the operation of data centers. It is instructive to note that a 1994 survey of 60 commercial buildings found that more than 50% had control system problems, 40% had problems with HVAC, more than 33% had faulty sensors, and about a 25% of them had energy management controls systems, economizers, and VSDs that did not operate properly (Pette 1996). The precise impact of commissioning, maintenance, and operations on energy efficiency and TCO is likely to be site specific, and to vary from site to site. It is, however, likely to impact almost all supporting aspects of the datacom facility, including HVAC cooling, HVAC fans, and electrical infrastructure.

Most mechanical, electrical, and HVAC faults in facilities can be traced to one of the following sources:

- faulty design
- wrong specification
- faulty installation
- faulty operation
- lack of maintenance

The first three sources are related to the initial design and construction, and can be avoided through proper commissioning. Adopting maintenance best practices can mitigate the last two sources.

All of the accepted best practices adopted in the construction, running, and maintenance of highly engineered process plants and commercial buildings are perhaps even more critical in the construction and operation of mission critical computing facilities. This appendix describes some of these best practices in the context of the ASHRAE Class 1 and Class 2 data centers. It is broadly divided into two sections: the first section deals with practices to be followed in the construction and hand-off of new facilities—or retrofitting of existing ones—in order to ensure trouble-free ownership and operation; the second section describes maintenance best practices that should be adopted to sustain trouble-free operation.

COMMISSIONING OF MISSION-CRITICAL FACILITIES

Commissioning is a systematic process of ensuring the eventual realization of a given design intent. Proper commissioning ensures certainty of intended outcome. In the case of critical facilities and support systems, proper commissioning would start from the design stage and follow through to hand-off from contractor to owner/operator. This applies to initial designs as well as specification and installation of additional equipment, or complete retrofitting of an existing facility. Recently, a number of practitioners have also advanced the idea of "continuous commissioning" and "retro-commissioning," which are intended to cover the usage and maintenance of a facility. However, the principles and practices espoused in these ideas are nothing more than the well-established principles of preventive and predictive maintenance, both of which will be covered in the next section.

To be most effective, commissioning should be approached as an independent verification process. Commissioning should provide an oversight function for the work performed in each stage of building a critical facility. This ensures that each stage of project implementation achieves the desired outcome. The American Institute of Architects' *Best Practices* identifies four phases to the commissioning of high performance buildings: predesign, design, construction, and warranty (AIA 2003).

In the predesign commissioning phase, the owner meets with the designers and architects to discuss the intended usage of a facility, including the criteria and benchmarks that will be used to judge the success or failure of implementation. Procedures for design development and documentation, schedules, roadmaps, and milestones are discussed at this stage. In a critical facility, the intended usage is to house datacom equipment, and the primary criterion for success is ensuring that the equipment continues to function with minimal or no breakdown. Thus, predesign commissioning must include specifications (primarily in terms of power and cooling requirements) for the housed equipment, and the criticality or class of intended usage; the latter determines the overall reliability goal for both the datacom equipment and building support infrastructure. Since datacom equipment refreshes frequently throughout the usage life of a facility, vision is required to accurately forecast future equipment requirements. Technology roadmaps, such as *Datacom Equipment Power Trends and Cooling Applications* (ASHRAE 2005c) are invaluable to this analysis.

As part of the design process, commissioning provides a framework for ensuring that the usage intent is not only met by the design, but also that downstream construction phase is implemented properly so as to not compromise any aspect of the design. This framework enforces a clear documentation of the original design intent, of assumptions and factors of safety built into the design, of criteria used in selecting equipment to provide particular functions, of mitigating factors for handling likely failure scenarios or events that might push the design envelope, etc. An independent simulation of normal operation and "what-if" scenarios must be a required component of design commissioning for critical facilities. This ensures that problems are identified and corrected early in the project implementation even before construction commences.

Commissioning in the construction phase involves procedural oversight to ensure that best practices are followed in the building construction and equipment installation. Functional tests of installed mechanical and electrical equipment are carried out to ensure that the equipment, as installed, meets the performance criteria used in selecting the equipment during the design phase.

Commissioning is traditionally associated with the hand-off phase of large construction projects. Hand-off of critical projects from contractor to owner/operators commonly involves a third-party commissioning provider. Commissioning of a new facility primarily covers start-up and functional testing of the installation to ensure that it meets the original design intent. This is normally an elaborate and extensive exercise in which all of the possible operating modes of the facility as a whole are tested, and major mechanical and electrical equipment may be tested individually against abnormal operating conditions. The commissioning provider records all observations and measurements taken during the start-up, including the performance and condition-indicating parameters of all major mechanical and electrical equipment. Once all the kinks have been worked out and all parties involved— the commissioner, the owner/operator, and the contractor—agree that the start-up performance meets the design intent, a warranty document may be issued against the documented performance data and the contractor may then sign off. The documented performance data serve as an as-new time signature upon which a proper preventive or predictive maintenance program may be built. This is the subject of the next section.

COMMISSIONING OF CONTROL SYSTEMS

This section looks at the need to integrate the commissioning strategy into the control system. In order for a system to operate at its peak energy efficiency, it needs to be controlled effectively. The control system coordination during the entire process needs to be managed from the beginning.

Background

Design professionals know that a faulty control system can seriously impair their desire to operate a reliable and efficient plant. Unfortunately, controls vendors often tend to work independently and in the "background." Controls are treated as a necessary but daunting task, understood by few. When the controls vendor is selected, the review of their submittals can have more emphasis on schedule and coordination with the mechanical contractor for delivery of dampers and valves, for example, than on system performance. In too many cases, the result is commissioning of the control system at the end of the project that is not on schedule, and pressure mounts to just "get it done."

It is important to open up the control system design process in a way that permits the design professional to incorporate the controls vendor into the commissioning process and maintain an understandable dialog.

System Performance

When selecting the mechanical equipment for a facility, typically the design professional will consider reliability and energy performance of the equipment. Since data centers are inherently large consumers of electricity, any and all strategies for optimizing energy efficiency are sought in equipment selection and real-time control and operation.

As an example, all chillers have an optimum performance range. Tight coordination between the chiller vendor and the controls vendor is necessary to operate the controls to maintain the right mix of chillers, allowing for peak performance. Strategies envisioned during design need to be confirmed and fine tuned during the commissioning process in the final system functional testing.

Project Factory Witness Test of Mechanical Controls

In data centers, factory witness testing is performed for the high-end critical support equipment. This is true of the generators, paralleling gear, UPS systems, and chillers. The controls for the paralleling gear in particular are tested. So why should we consider this for the mechanical system controls?

As a part of the process of designing the right sequence, issues are discussed and adjustments can be made to sequences during the construction phase of the project. The frustration develops when the system is being commissioned while the control vendor is programming. In an effort to meet critical turnover deadlines, the time allotted for commissioning may be reduced, commissioning for reliability becomes the primary concern, and energy efficiency can end up taking a back seat.

By requiring a control vendor to perform a factory witness test of his system, critical reliability issues can be addressed at the 90% level. This allows the design professional more time to operate his design and fine tune for the energy efficiency expected from the system. The ability to perform extra tests under load is increased, and the energy-efficient attributes of the design do not become a victim of the schedule.

MAINTENANCE OF MISSION-CRITICAL FACILITIES

There are, in essence, three basic types of maintenance programs: failure-based maintenance, planned preventive maintenance, and maintenance based on equipment condition monitoring (predictive maintenance).

Failure-based maintenance is a reactive program that waits for an actual failure to occur before taking a corrective action. Although this may require the least amount of planning and upfront costs (manpower and spare parts) among all three types of programs, it permits unplanned interruptions in facility operations and is therefore not suitable for critical facilities that are required to maintain a 24/7 year-round uptime. Unfortunately, though, this is the most common program in place at most sites, especially since most small- to medium-size facilities are piggy-backed on existing buildings that are maintained by separate facility operations units. Failure-based maintenance allows for catastrophic events, which can lead to prolonged and expensive downtimes or even loss of an entire facility.

A preventive maintenance program follows manufacturer-recommended maintenance schedules for each piece of equipment. Although this may lead to less frequent equipment breakdowns, it can also result in excessive and unnecessary maintenance activity with each piece of equipment being on a different maintenance schedule. The cost for maintaining such a program in a large facility can become prohibitive and unjustifiable, especially since it can still be ineffective in preventing actual breakdowns and interruption of service.

Clearly a critical facility that is required to maintain 24/7 year-round uptime will benefit from a program that can predict the onset of equipment deterioration and timely fixing of potential problems before they can lead to actual breakdowns. Predictive maintenance, which is centered on sustaining high equipment reliability, is such a program. Predictive maintenance is a proactive program in which the health or operating conditions of critical equipment are continuously monitored in order to predict the onset of equipment deterioration. Equipment condition monitoring enables the facility operator to ward off untimely breakdowns by detecting potential problems and fixing them before they can occur. The nature of this program makes it uniquely suitable for implementation in a critical computing facility. Depending on the level of implementation (of condition monitoring), this type of program can also be the least expensive maintenance program since maintenance is on-demand and carried out as necessary. Condition monitoring also allows for the development of an extensive knowledge base on a given facility. This knowledge base may be used not only to take preventive maintenance actions, but also for training new personnel on the operation and peculiar characteristics of a facility.

Condition Monitoring (CM)

The bedrock of reliability-centered maintenance is dedicated equipment health monitoring. In this strategy, the condition-indicating parameters of all or critical pieces of equipment are continuously monitored and compared against a baseline signature. Ideally, the baseline signature would be the as-new performance data recorded during commissioning and hand-off of the facility. In cases where these data do not exist, a "re-commissioning" project (e.g., energy audit) may be carried out to record and evaluate the performance of the plant as well as individual pieces of equipment. Re-commissioning is very useful for finding where inefficiencies exist and where upgrades to latest technology may be beneficial to the overall efficiency of the plant. When new equipment is installed, the baseline signature should be updated with the as-new performance data of the equipment.

Types of CM. CM techniques may be broadly classified into three main types:

1. *Inspection* involves qualitative tests using human senses (touch, site, smell, and hearing), and simple tools such as oil gauge or liquid level sight glass.
2. Occasional *condition checks* that involve measuring parameters to compare against a baseline signature or pre-determined design value.
3. *Trend monitoring* involves historical tracking of the condition indicating parameter. This is the normally the most effective CM type, and the most useful for developing a comprehensive knowledge base about the operation of the facility plant.

CM Metrics. CM is a very useful and practical technique for predicting if a failure is likely to occur in a piece of equipment. Since CM is done online while the equipment is in operation, it is not necessary to disrupt the operation of the facility and, thus, is a particularly suitable maintenance program to adopt in mission critical facilities. When properly applied, CM is a very effective tool to measure deterioration in a piece of equipment or system. CM can be used to measure the extent of deterioration, the rate of deterioration, and the mean time-to-failure of a piece of equipment.

Effectiveness of a CM Program. The main factors governing the effectiveness of a condition monitoring program include the following:

1. *Parameters Used to Indicate the Operating Condition of the Equipment.* The set of parameters selected to monitor should be those that are easily quantifiable and measurable. Among all parameters that are measurable, the selection should be biased toward those whose measurement techniques can be automated in a data acquisition system. Usually there are one or two parameters that are dominant indicators of the health or operating efficiency of the

equipment. Obviously, if the output of the equipment (i.e., the main reason for using the equipment) is easily measurable, then it is logical to select this output as the condition indicating parameter. However, in a complicated piece of machinery or system, the output of the system is the result of several processes or components working together. In such cases, measuring the output, while useful as an indication of the overall health of the machine, may not be so useful in a diagnostic effort to pinpoint the source of the problem. Thus, it is often useful to instrument other parameters (e.g., surface temperatures, oil temperature, etc.) on the pieces that make up the equipment.

2. *CM Techniques are Employed.* Selecting the appropriate technique or combination of techniques for a given program is crucial to the success or failure of the maintenance program. The "best" technique can vary from machine to machine and situation to situation. Although trend monitoring is the most sophisticated and most rigorous, it is easy to get buried in a sea of data and miss the point of the health check. It is not necessary or effective to implement a sophisticated data acquisition program when a simple visual inspection can suffice. Normally a typical critical facility would require a combination of techniques given the mix of equipment that exist.

3. *Level of Automation.* The level of automation of CM typically determines the effectiveness of the maintenance program. The type and mix of equipment in the cooling plant, electrical power distribution system, and other support infrastructure of a critical facility demands a high level of automation in an effective CM program. Note that a successful automation system deals with not only automatic data acquisition, but also management and assessment of the collected data. There are several commercial supervisory control and data acquisition (SCADA) and building management (BM) systems available that do a good job of collecting and archiving instrumentation data. However, most of these systems do a poor job of assessing the data for equipment and system health, or none at all, and, thus, they must be used in combination with computerized maintenance management (CMM) systems to implement an effective CM program. Note that in most cases commercial off-the-shelf CMM systems must be customized to the specific installation to be used effectively.

SUMMARY

Quality commissioning, maintenance, and operations are all very important to the energy efficiency and TCO of a datacom facility. The following are particularly important points:

• Commissioning should start with the predesign phase of the building design, so that energy efficiency attributes can be designed into the system.

- Controls commissioning is a particularly important part of the commissioning process, allowing for system energy performance to be documented, tested, and fine tuned prior to turning the system turn over to the owner.
- A quality maintenance program is one of the best ways to maintain equipment operation at high efficiency, achieve high reliability, and, thus, optimize facility TCO. A predictive maintenance program with condition monitoring is highly recommended.

Appendix C

Telecom Facility Experiences

INTRODUCTION

This appendix looks at energy-efficiency measures in facilities where communications equipment is located. Central offices (COs) are used to house mainly telecom equipment to route telephone calls to the correct addresses, but modern installations contain both data and telecom equipment. Since a large amount of research and practical experiences are available, this appendix is by no means an exhaustive review of the matter. The goal here is to provide a high level view of some typical telecom facility trends and challenges that might trigger interest in learning more about some of the energy-saving techniques.

Background

Today's COs house a combination of voice, data, and video communications equipment. These facilities are nodes of the network, providing functionalities such as switching, routing, processing, transport, regeneration, and storage. There is a trend toward a converged communications platform for transport and manipulation of digital bits regardless of source or origin.

During the early days of telecom, electromechanical telecom equipment was bolted directly to the concrete slab, and ventilation air was provided from the top of the room. Cooling, temperature, and humidity requirements were initially limited. When digital equipment was introduced, the regulated telecom monopoly drove the standards development for nearly anything associated with its business, including environmental specifications. Equipment robustness was at the forefront.

Telecom Equipment

Greater performance, functionality, and processor speeds mean greater energy use and, thus, greater heat load into the equipment room. Higher heat loads necessitate more air conditioning for maintaining room conditions at specified levels. There is this cycle of deploying new technologies and then having to cope

201

with their greater energy needs, thermal management challenges, and building air-conditioning demands.

According to industry roadmaps (NEMI 2000, 2002, 2004), telecom equipment located in COs leads in terms of power (W), power/space density (W/ft^3 [W/m^3]), and power/footprint density (W/ft^2 [W/m^2]). Telecom equipment energy is consumed by providing the desired functionality, and the vast majority of the energy is dissipated as heat into the space housing the equipment requiring subsequent cooling.

Current high-traffic COs contain both electronic and optical equipment. An analysis of such facilities (Vukovic 2004, 2005) has shown that for every gigabit per second (Gbps) of data transfer, 12W to 80W are dissipated depending on type of equipment, function, and vendor. An efficient CO might have an average demand of 50 W/Gbps.

Since terabit-per-second traffic is not uncommon, heat dissipation in excess of 300 kW can be expected in large installations (Vukovic 2003, 2005). Consequently, service providers who manage high-traffic facilities are expressing concerns, particularly for the situation when they have to add new high-powered equipment. High-power requirements not only pose concerns for power systems, but also for cooling systems.

Results from modeling studies (Herrlin 1996) indicate that the economic benefits of reducing the energy consumption by telecom equipment are greater than those of any other single energy-efficiency measure. However, trying to get a handle on moving equipment toward greater efficiency is difficult without recognized industry metrics. Even if metrics were established, requirements may not be far behind. Could such requirements stifle development of new equipment technologies? Thermal issues at the chip level, as well as rising energy prices, have become powerful incentives for improved electronic equipment energy efficiency.

Telecom service providers do not generally emphasize energy efficiency as they review new products for deployment; the focus is to get new services to customers as quickly as possible. However, energy usage at the equipment level will become more important when service providers realize a significant increase in the cost of the next deployment. In any event, end users would welcome reductions in energy usage. The rest of this chapter reviews a number of techniques for reducing the need for air-conditioning equipment and energy in COs.

Environmental Specifications

Temperature, humidity, and noise specifications for COs differ from those used in data centers. One key reason is that telecom standards were driven by large regulated end users, whereas mainly equipment manufacturers developed data center guidelines. Since end users emphasized equipment robustness (through the NEBS testing requirements), the equipment is allowed to operate in a wider range of environmental conditions.

Table C.1   Comparison between Class 1 & 2 Data Centers and NEBS

(@ Rack Air Intake)	Recommended (Facility)	Allowable (Equipment)
Temperature Data Centers Telecom NEBS	64.4°F–80.6°F (18°C–75°C) 65°F–80°F (18°C–27°C)	59°F–90°F (15°C–32°C) 41°F–104°F (5°C–40°C)
Humidity (RH) Data Centers Telecom NEBS	41.9°F dew point–60% RH and 59°F dew point; Max 55%	20%–80% 5%–85%

Table C.1 shows the differences between data centers (ASHRAE 2009) and NEBS (Telcordia 2001, 2002). Energy savings associated with appropriate environmental controls have been studied by Telcordia (Bellcore) (Herrlin 1996). Generally, increasing the temperature range in which telecom equipment operates is cost-effective since it decreases the energy costs of operating the HVAC system. Data centers maintain an indoor relative humidity above a given setpoint to reduce equipment failure due to electrostatic discharge (ESD). Analyses of telecom facilities (Herrlin 1996) suggest, however, that humidification is generally not a cost-effective solution to ESD failures. Other solutions are recommended, notably personal grounding. Partly due to these findings, most COs do not use active humidification.

Airflow Management

The standard and preferred way of introducing cool air into COs is through overhead ducted air diffusers (VOH) and using hot and cold aisles. Various issues related to airflow and thermal management in COs are documented in NEBS GR-3028-CORE (Telcordia 2001) based on a broad industry collaboration with 20 major equipment vendors and end users.

More recent research utilizing the RCI suggests that over-head cooling be a highly effective and efficient way of cooling equipment rooms (Herrlin 2005a; Herrlin and Belady 2006). A number of new data centers have also adopted this air-distribution protocol in combination with an efficient central HVAC plant, rather than conventional underfloor cooling (VUF) with CRAC units.

Central Air Handlers

The standard way of cooling COs is to use large central air handlers and duct-conditioned air to the equipment space. By using large centralized mechanical equipment, very few and energy-efficient components can be specified. Maintenance is simplified by the central location of large units. These observations are in stark contrast to CRAC systems deployed in most data centers. Multiple units with a large number of components and lack of coordination often lead to infighting. For example, it is not uncommon that one unit is humidifying while another is dehumidifying.

Air-Side Economizers

The majority of COs operate successfully with sensible or enthalpy economizers (Herrlin 2005b). With the central air handlers, air-side economizers are relatively easy to incorporate. Although the energy savings can be significant, economizers are highly climate dependent. Since the indoor conditions in telecom central offices are often allowed to "float," economizers can effectively be utilized in many climates. Specifically, the generally low need for humidification makes air-side economizers attractive also in dry regions.

Nevertheless, there are a number of issues that need to be addressed when introducing large volumes of outdoor air into equipment rooms, e.g., air filtration. The more outdoor air, the more fine particles are brought in from the outside. This is especially worrisome in urban areas. Although the equipment turnover rate is not as frantic as in data centers, the shorter turnover rate of equipment eases the soiling concerns. Since high cooling loads result in high air exchange rates and large filter banks, air-side economizers may become impractical.

Equipment with an expected turnover rate of more than five years may provide sufficient time to accumulate enough fine particles to jeopardize the equipment reliability should high indoor humidity be encountered. Hygroscopic dust failures (HDFs) have the grave characteristics that it often results in multiple unrelated failures. To prevent such events, effective air filtration is the first line of defense. In addition, dust storms in desert areas have been reported to wipe out filter banks completely.

A documented effect of air-side economizer operation is indoor humidity and temperature swings due to rapid weather changes or while starting/stopping mechanical cooling. Another observation is that some economizers have been disconnected in humid and/or hot locations. This could be due to inappropriate design (no override or lock-out controls), poor staff training (poor understanding of proper operation), and/or maintenance issues.

Outdoor Air Ventilation

Minimum outdoor air ventilation is mainly deployed to dilute indoor generated organic compounds (VOCs) that could harm the electronics. The minimum air exchange rate is determined at the point when further increase of outdoor air produces diminishing returns. Telecom practice is a minimum of 0.25 ach (Weschler and Shields 1998). Beyond that point, HVAC energy is simply wasted by conditioning additional outdoor air. Of course, during air-side economizer operation, the air exchange rate is well above this threshold. The minimum outdoor damper setting of the economizer should be near the flow rate corresponding to the minimum air-exchange rate.

Space Pressurization

Traditionally, in telecom applications, space pressurization was deployed to keep out submicron particles from electronic equipment rooms by limiting air infiltration though the building envelope. Since these particles are mainly generated by various combustion (e.g., automobiles) processes, they originate from the outdoors. However, research indicates that space pressurization could be counterproductive in COs (Herrlin 1997). In addition, pressurization is associated with an energy penalty, especially in leaky buildings.

Pressurization becomes more useful in tight buildings with effective air filtration. The current building stock of COs is aging, and many of these buildings are leaky due to a number of reasons. One is gravity-controlled exhaust dampers (for controlling the indoor-outdoor pressure differential) that are notoriously leaky. Another reason is that central office buildings often are free-standing "boxes" with five external surfaces. A third reason is that maintenance routines may be poor such as keeping external doors closed.

TCO

Most of the items discussed above affect the TCO. Money saved by reducing the energy use improves the company's bottom line. Although energy efficiency may not always be a top priority, it is often a welcome "side-effect" of an intelligent CO design providing high network reliability. As the energy costs are expected to continue to climb in the years to come, the priorities may shift. A general discussion on TCO can be found in Chapter 10 of this publication.

Appendix D

Applicable Codes, Standards, and Other Organizations

CODES

International Codes

The International Building Code (IBC) primarily relies on the latest version of ASHRAE Standard 90.1 (ASHRAE 2004a). The actual version that is applicable to a specific state or country is variable, and is dependent on the date of adoption for each state. In addition, states or countries are free to adopt only a portion of ASHRAE Standard 90.1, and to make modifications for other versions.

National Codes

There are several energy codes that may pertain to data center energy efficiency. The standard with the greatest impact is ASHRAE Standard 90.1.

State and Local Codes

Various states have energy codes that may be more restrictive (i.e., require higher levels of energy efficiency) than ASHRAE Standard 90.1.

The highest-profile state energy code is probably the California Energy Code http://www.energy.ca.gov/title24/.

STANDARDS

ASHRAE Standard 90.1 is the standard that most international, national, state, and local codes use as a baseline. This standard is under "continuous maintenance," meaning that it can be changed on an annual basis to stay abreast of current technology, rather than being amended at a fixed interval. ASHRAE currently maintains a link to a read-only version of this standard on its Web site at no charge. The link address (at the time of publication) to the 2004 version of this standard is: http://www.realread.com/prst/pageview/browse.cgi?book=1931862664.

The rating of CRACs is governed by ASHRAE Standard 127 (latest version) (ASHRAE 2007b).

OTHER ORGANIZATIONS

Several other organizations have expressed interest in data center energy efficiency. These include government agencies, utilities, for-profit companies, and nonprofit organizations. Select organizations that offer additional information relating to data center energy efficiency are as follows:

Alliance to Save Energy http://www.ase.org/
Lawrence Berkeley National Laboratory http://www.lbl.gov/
Pacific Gas and Electric Company: http://www.pge.com/
Rocky Mountain Institute: http://www.rmi.org/
Standard Performance Evaluation Corporation http://www.spec.org/
The Green 500 List: http://www.green500.org/Home.html
The Green Grid http://www.thegreengrid.org/home
The Uptime Institute http://www.uptimeinstitute.org/
United States Department of Energy http://www.energy.gov/
United States Environmental Protection Agency http://www.epa.gov/

ASHRAE also has a listing of codes and standards published by various societies and associations. Included is a list of energy codes and standards, and standards relating to testing and rating of specific types of HVAC equipment, such as air conditioners, air filters, chillers, coils, compressors, condensers, condensing units, cooling towers, dehumidifiers, fans, heat exchangers, humidifiers, motors and generators, pumps, refrigerants, and thermal storage (ASHRAE 2007h).

Appendix E

Sample Control Sequences

This appendix provides some sample detailed control sequences. The sequences provided are for air-side economizer control, and for air-side wet-bulb economizer control.

AIR-SIDE ECONOMIZER SEQUENCES

Control Sequence

1. **When Outdoor Dry-Bulb Temperatures are below the Required Data Center Supply Air Temperature Setpoint.** Outdoor air and return air dampers shall modulate to satisfy the data center supply air temperature setpoint. Building air delivery temperature sensors shall be located downstream of the supply air fan and steam humidification device to account for any sensible heat added by these components. Steam humidification shall be added to maintain the required room air dew-point condition. Return air or relief air fans shall have fan-speed control to maintain the required data center positive pressure. Variable-air-volume (VAV) supply air systems may require a minimum fresh air injection fan to provide code ventilation for building occupants in compliance with ASHRAE Standard 62 during extreme cold ambient conditions.
2. **When Outdoor Air Dry-Bulb Temperatures are above the Required Supply Air Temperature Setpoint, but the Wet Bulb (Enthalpy) is below the Return Air Wet Bulb (Enthalpy).** Outdoor air dampers shall be 100% open and return air dampers closed. The unit-mounted cooling coil valve shall open to maintain the required supply air dry-bulb setpoint temperature. Steam humidification shall be added as required to maintain the room dew-point temperature. Room positive pressure shall be maintained by speed control of the return or relief air fans.

Note: If the air is not being cooled to saturation, and the COP of the conditioning process is different for outdoor air vs. return air (which may be the case for high supply air temperatures where outdoor air must be dehumidified but return air does not), the energy used in the conditioning process needs to be calculated to determine the most energy-efficient control strategy. A simple enthalpy comparison is not sufficient in this case.

3. **When Outdoor Air Wet Bulb (Enthalpy) is Higher than the Building Return Air Wet Bulb (Enthalpy).** Outdoor air dampers shall close to their minimum position, as required for either ventilation or room pressurization, and return air dampers shall be 100% open. For dry climates, steam humidification may be required to maintain room dew-point conditions. Unit-mounted cooling coil valves shall modulate to deliver the required building supply air dry bulb temperature setpoint. Return or relief fan speed shall be modulated to maintain the required room positive pressure.

AIR-SIDE WET-BULB ECONOMIZER SEQUENCES

When a central station air-handling unit is furnished with an adiabatic cooling component, this device will provide free humidification during cold ambient conditions using the heat generated within the data center.

During more moderate outdoor air temperatures, the direct evaporative cooling system will greatly extend the system free cooling ton-hours especially in more arid climates. These are the ambient economizer hours when outdoor dry-bulb temperatures are above the building supply air setpoint but the wet-bulb temperature is less than the room required maximum dew-point temperature. In dry western climates, it is not unusual for there to be 25% to 40% of the annual ambient hourly conditions that reside within this psychrometric chart envelope. These are all hours when an air-side economizer, without evaporative cooling, would require both mechanical cooling and humidification.

Positioning of the cooling coil downstream of the adiabatic device will provide the lowest refrigeration tonnage requirement, since the coil leaving air condition will be at saturation, thereby providing a very good data center dew-point control.

An additional benefit of the direct evaporative cooling component is the air scrubbing function it performs. Larger contaminants are removed from the supply air, providing backup protection for data center electronic equipment. A ridged media-type evaporative cooler should be used to ensure that nothing but water vapor is released into the supply airstream. For more information on evaporative cooling and humidification, please review Chapter 19 of the *2004 ASHRAE Handbook—HVAC Systems and Equipment* and Chapter 51 of the *2007 ASHRAE Handbook—HVAC Applications.*

Control Sequence

1. **When Outdoor Air Wet-Bulb Temperature (Enthalpy) is Lower than the Data Center Dew-Point Temperature Setpoint.** The direct evaporative cool-

ing component recirculation water pump is on and outdoor air and recirculation air dampers shall modulate to provide a mixed air condition which, when supplied to the wet ridged media pad, produces the required data center dry-bulb setpoint temperature. Humidification will not be required if the saturation efficiency of the evaporative cooling media is selected at 90% or higher. (More exact control of supply air dew point may require face and bypass dampers for the adiabatic cooler/humidifier component.) Return and/or relief air fans shall modulate speed as required to maintain the specified data center pressurization. Winter room dew-point conditions may be reset lower in cold weather to reduce both supply air dry-bulb temperatures and room relative humidity. VAV supply air systems may require a minimum fresh air injection fan to provide code ventilation for building occupants during extreme cold ambient conditions.

2. **When Outdoor Air Wet-Bulb Temperatures (Enthalpy) are above the Data Center Dew-Point Temperature Setpoint but are below the Data Center Return Air Wet-Bulb Temperature (Enthalpy).** The evaporative cooling recirculation pump is on and the outdoor air damper is 100% open with the return air damper closed. The adiabatic device provides reduced cooling and free humidification in this mode for dry climates. For more humid climates, if ambient dew-point conditions are above the room dew point, the direct evaporative cooler pump may be turned off. Return air or relief air fans shall run at the speed as required to maintain the data center positive pressure. The cooling coil valve shall modulate open as required to provide the data center supply air dry-bulb temperature.

Note: If the air is not being cooled to saturation, and the COP of the conditioning process is different for outdoor air vs. return air (which may be the case for high supply air temperatures where outdoor air must be dehumidified but return air does not) the energy used in the conditioning process needs to be calculated to determine the most energy-efficient control strategy. A simple enthalpy comparison is not sufficient in this case.

3. **When Outdoor Air Wet-Bulb Temperature (Enthalpy) is Higher than the Data Center Return Air Wet-Bulb Temperature (Enthalpy).** The evaporative cooling recirculation pump is off and the outdoor air damper is set at its minimum position as required for either building pressure or ventilation. The cooling coil valve shall modulate open as required to provide the data center supply air dry-bulb temperature. Return air or relief air fans shall adjust their speed as required to maintain data center positive pressure with maximum recirculation of building return air. For very arid climates, the evaporative cooling recirculation pump may need to cycle on briefly to maintain room design dew-point conditions.

Appendix F

SI Units for Figures and Tables

Most of the figures and tables in this book are presented in I-P units. This appendix contains the corresponding figures and tables in SI units, unless they are available as such in other ASHRAE publications.

CHAPTER 2

Table 2.1 Class 1, Class 2, and NEBS Design Conditions

Condition	Class 1/Class 2		NEBS	
	Allowable Level	Recommended Level	Allowable Level	Recommended Level
Temperature control range	15°C–32°C[a,f] (Class 1) 10°C–35°C[a,f] (Class 2)	18°C–27°C[a,g]	5°C–40°C[c,f]	18°C–27°C[d]
Maximum temperature rate of change	5°C per hour[a]		1.6°C/min.[d]	
RH control range	20%–80% 17°C max dew point[a] (Class 1) 21°C max dew point[a] (Class 2)	41.9°F dew point–60% RH and 59°F dew point[a, g]	5%–85% 28°C max dew point[c]	Max 55%[e]
Filtration quality	65%, min 30%[b] (MERV 11, min MERV 8)[b]			

[a]These conditions are inlet conditions recommended in *Thermal Guidelines for Data Processing Environments* (ASHRAE 2009).
[b]Percentage values per the ASHRAE Standard 52.1 dust-spot efficiency test. MERV values per ASHRAE Standard 52.2. Refer to Table 8.4 of *Design Considerations for Datacom Equipment Centers* (ASHRAE 2005d) for the correspondence between MERV, ASHRAE 52.1, and ASHRAE 52.2 filtration standards.
[c]Telecordia 2002 GR-63-CORE.
[d]Telecordia 2001 GR-3028-CORE.
[e]Generally accepted telecom practice. Telecom central offices are not generally humidified, but grounding of personnel is common practice to reduce ESD.
[f]Refer to *Thermal Guidelines for Data Processing Environments* (ASHRAE 2009) for temperature derating with altitude.
[g]20°C/40% RH corresponds to a wet-bulb temperature of 12°C. *Caution:* operation of DX systems with a *return* wet-bulb temperature below 12°C has a likelihood of causing freezing coils.

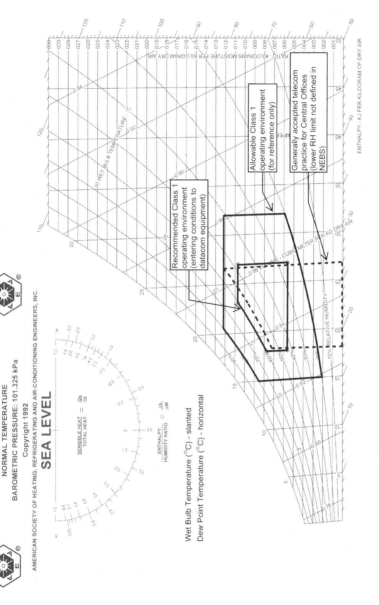

Figure 2.2a Recommended data center Class 1, Class2, and NEBS operating conditions (SI units).

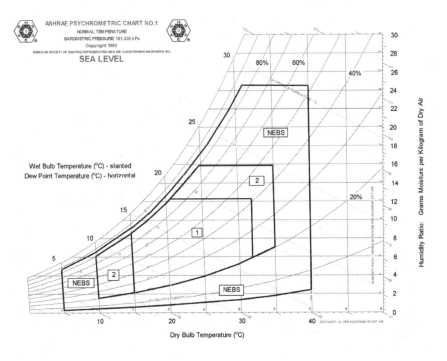

Figure 2.2b Allowable data center Class 1, Class 2 and NEBS operating conditions (SI units).

Metric Equivalent of Savings Example #1 (Section 2.3.2)

Savings Example #1: As an example of the cost savings possible from resetting the minimum relative humidity from 45% to 40%, consider the following:

Assume that a data center requires $16,990 \text{ m}^3/\text{h}$ of outdoor air, and that for 50% of the year the air needs to be humidified to either (a) 45% RH or (b) 40% RH at a temperature of 22.2°C. Electric source heat is used to provide the humidification, and the average energy cost is \$0.10/ kWh.

Enthalpy of 22.2°C, 45%RH air: 41.40 kJ/kg
Enthalpy of 22.2°C, 40%RH air: 39.26 kJ/kg
Difference: 2.14 kJ/kg
Mass flow rate: $16,990 \text{ m}^3/\text{h} \times 1.185 \text{ kg/m}^3 = 20,133 \text{ kg/h}$
Energy savings of 40% setpoint: $20,133 \text{ kg/h} \times 2.14 \text{ kJ/kg} = 43,085 \text{ kJ/h}$
$43,085 \text{ kJ/h} \times 0.0002778 \text{ kJ/h/ kW} = 11.97 \text{ kW}$
$11.97 \text{ kW} \times 4380 \text{ h/yr} \times \$0.10/ \text{kWh} = \$5243/\text{yr}$

CHAPTER 3

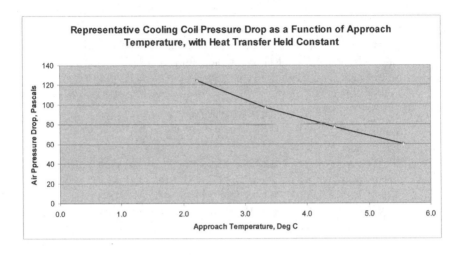

Figure 3.3   Representative cooling coil air pressure drop as a function of approach temperature between entering chilled-water temperature and leaving air temperature.

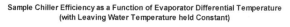

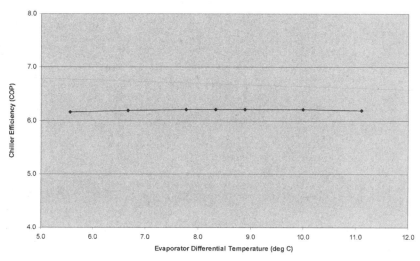

Figure 3.6 Sample chiller efficiency as a function of evaporator differential temperature (with other parameters held essentially constant).

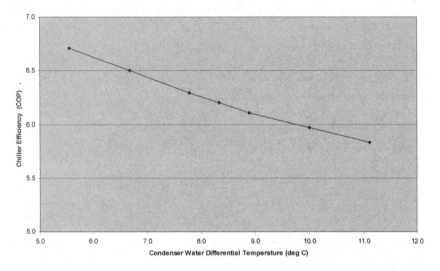

Figure 3.7 Sample chiller efficiency as a function of condenser-water differential temperature (with other parameters held essentially constant).

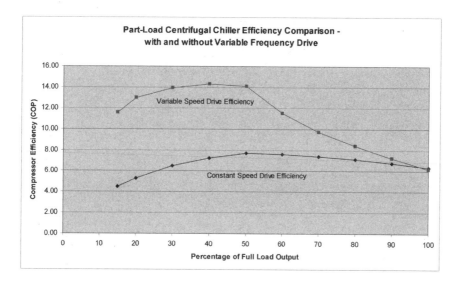

Figure 3.8 Sample part-load centrifugal chiller efficiency with and without variable frequency drive.

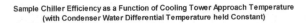

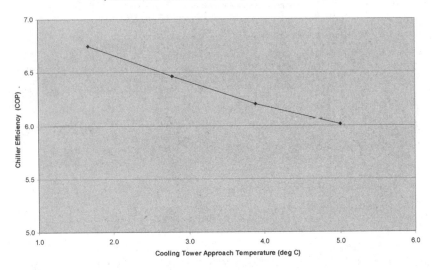

Figure 3.10 Sample chiller efficiency as a function of cooling tower approach temperature (with other parameters held essentially constant).

CHAPTER 9

Table 9.4 Sample Comparison of Transport COP for Air, Water, and Phase-Change Refrigerant Heat Rejection

	Air	Water	R-134a
Density	1.217 kg/m^3 (@20C)	998 kg/m^3	1206 kg/m^3
Specific heat	1026 J/kg·K	4187 J/kg·K	215,900 J/k (heat of vapor)
Volumetric heat capacity	1249 J/m^3·K	4,178,426 J/m^3·K	260,375,400 J/m^3·K
		3346 ratio	N/A
Volumetric heat capacity	0.01862 Btu/ft^3·°F	62.3 Btu/ft^3·°F	6988 Btu/ft^3
Typical heat rise	12.2 K	6.67 K	Phase change
Volumetric heat-transfer content	15,263 J/m^3	27,856,176 J/m^3	260,375,400 J/m^3
		1825 ratio to air	17,060 ratio to air
Flow rate	1.00 m^3/s	1.00 m^3/s	1.00 m^3/s
Heat transfer per unit time	15,263 J/s	27,856,176 J/s	260,375,400 J/s
Energy removed per unit time	15,263 W	27,856,176 W	260,375,400 W
Typical HVAC system pressure drop	623 Pa	149,453 Pa	180,602 Pa
		240 ratio	290 ratio to air
Required transport power per unit flow rate	623 W/m^3/s	149,453 W/m^3/s	180,602 W/m^3/s
COP (energy removed/input prime mover energy input)	24.5 W/W	186.4 W/W	1442 W/W
Ratio (liquid:air)		7.60 ratio to air	58.8 ratio to air
Ratio (phase-change refrigerant:water)			7.74 ratio to water

Appendix G

2008 ASHRAE Environmental Guidelines for Datacom Equipment— Expanding the Recommended Environmental Envelope[1]

The recommended environmental envelope for IT equipment is listed in Table 2.1 of ASHRAE's *Thermal Guidelines for Data Processing Environments* (2004). These recommended conditions, as well as the allowable conditions, refer to the inlet air entering the datacom equipment. Specifically, it lists for data centers in ASHRAE Classes 1 and 2 (refer to *Thermal Guidelines* for details on data center type, altitude, recommended vs. allowable, etc.) a recommended environment range of 20°C to 25°C (68°F to 77°F) (dry-bulb temperature) and a relative humidity (RH) range of 40% to 55%. (See the allowable and recommended envelopes for Class 1 in Figure G.1 below.)

To provide greater flexibility in facility operations, particularly with the goal of reduced energy consumption in data centers, ASHRAE Technical Committee (TC) 9.9 has undergone an effort to revisit these recommended equipment environmental specifications, specifically the recommended envelope for Classes 1 and 2 (the recommended envelope is the same for both of these environmental classes). The result of this effort, detailed in this appendix, is to expand the recommended operating environment envelope. The purpose of the recommended envelope is to give guidance to data center operators on maintaining high reliability and also operating their data centers in the most energy-efficient manner. The allowable envelope is where IT manufacturers test their equipment in order to verify that it will function within those environmental boundaries. Typically, manufacturers perform a number of tests prior to the announcement of a product to verify that it meets all the functionality requirements within this environmental envelope. This is not a statement of reliability but one of functionality of the IT equipment. However, the recommended envelope is a statement of reliability. IT manufacturers recommend that data center operators maintain their environment within the recommended envelope for extended periods of time. Exceeding the recommended limits for short periods of

---

1. This appendix modifies the first edition of *Thermal Guidelines for Data Processing Environments* (ASHRAE 2004). It is included here for reference only.

time should not be a problem, but running near the allowable limits for months could result in increased reliability issues. In reviewing the available data from a number of IT manufacturers, the 2008 expanded recommended environmental envelope is the agreed-upon envelope that is acceptable to all IT manufacturers, and operation within this envelope will not compromise overall reliability of IT equipment. The previous and 2008 recommended envelope data are shown in Table G.1.

Neither the 2004 nor the 2008 recommended operating environments ensure that the data center is operating at optimum energy efficiency. Depending on the cooling system, design, and outdoor environmental conditions, there will be varying degrees of efficiency within the recommended zone. For instance, when the ambient temperature in the data center is raised, the thermal management algorithms within some datacom equipment increase the speeds of air-moving devices to compensate for the higher inlet air temperatures, potentially offsetting the gains in energy efficiency due to the higher ambient temperature. It is incumbent upon each data center operator to review and determine, with appropriate engineering expertise, the ideal operating point for their system. This will include taking into account the recommended range and site-specific conditions. Using the full recommended envelope is not the most energy-efficient environment when a refrigeration cooling process is being used. For example, the high dew point at the upper areas of the envelope result in latent cooling (condensation) on refrigerated coils, especially in direct expansion (DX) units. Latent cooling decreases the available sensible cooling capacity for the cooling system and, in many cases, leads to the need to humidify to replace moisture removed from the air.

The ranges included in this document apply to the inlets of all equipment in the data center (except where IT manufacturers specify other ranges). Attention is needed to make sure the appropriate inlet conditions are achieved for the top portion of IT equipment racks. The inlet air temperature in many data centers tends to be warmer at the top portion of racks, particularly if the warm rack exhaust air does not have a direct return path to the computer room air conditioners (CRACs). This warmer air also affects the RH, resulting in lower values at the top portion of the rack.

Table G.1   Comparison of 2004 and 2008 Recommended Environmental Envelope Data

	2004	2008
Low-End Temperature	20°C (68°F)	18°C (64.4°F)
High-End Temperature	25°C (77°F)	27°C (80.6°F)
Low-End Moisture	40% RH	5.5°C DP (41.9°F)
High-End Moisture	55% RH	69% RH and 15°C (59°F DP)

The air temperature generally follows a horizontal line on the psychrometric chart, where the absolute humidity remains constant but the RH decreases.

Finally, it should be noted that the 2008 change to the recommended upper temperature limit from 25°C to 27°C (77 °F to 80.6 °F) can have detrimental effects on acoustical noise levels in the data center. See the section "Acoustical Noise Levels" later in this appendix for a discussion of these effects.

The 2008 recommended environmental envelope is shown in Figure G.1. The reasoning behind the selection of the boundaries of this envelope are described below.

DRY-BULB TEMPERATURE LIMITS

Part of the rationale in choosing the new low- and high-temperature limits stemmed from the generally accepted practice for the telecommunication industry's central office, based on NEBS GR-3028-CORE (Telcordia 2001), which uses the same dry-bulb temperature limits as specified here. In addition, this choice provides a precedence for reliable operation of telecommunication electronic equipment based on a long history of central office installations all over the world.

Low End

From an IT point of view, there is no concern in moving the lower recommended limit for dry-bulb temperature from 20°C to 18°C (68°F to 64.4°F). In equipment with constant-speed air-moving devices, a facility temperature drop of 2°C (3.6°F) results in about a 2°C (3.6°F) drop in all component temperatures. Even if variable-speed air-moving devices are deployed, typically no change in speed occurs in this temperature range, so component temperatures again experience a 2°C (3.6°F) drop. One reason for lowering the recommended temperature is to extend the control range of economized systems by not requiring a mixing of hot return air to maintain the previous 20°C (68°F) recommended limit. The lower limit should not be interpreted as a recommendation to reduce operating temperatures, as this could increase hours of chiller operation and increase energy use. A noneconomizer-based cooling system running at 18°C (64.4°F) will most likely carry an energy penalty. (One reason to use a noneconomizer-based cooling system would be a wide range of inlet rack temperatures due to poor airflow management; however, fixing the airflow would likely be a good first step toward reducing energy.) Where the setpoint for the room temperature is taken at the return to cooling units, the recommended range should not be applied directly, as this could drive energy costs higher from overcooling the space. The recommended range is intended for the inlet to the IT equipment. If the recommended range is used as a return air setpoint, the lower end of the range (18°C to 20°C) (64.4°F to 68°F) increases the risk of freezing the coils in a DX cooling system.

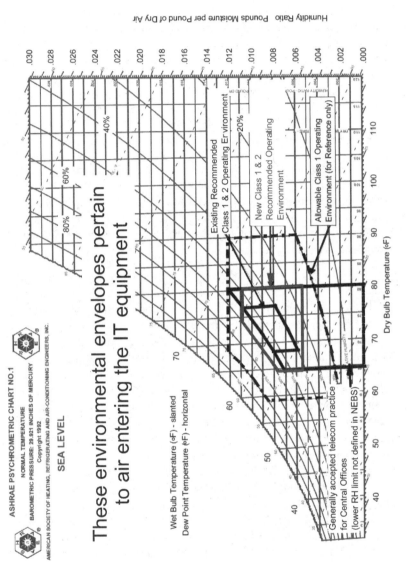

Figure G.1 2008 recommended environmental envelope (new Class 1 and 2).

High End

The greatest justification for increasing high-side temperature is to increase hours of economizer use per year. For noneconomizer systems, there may be an energy benefit by increasing the supply air or chilled-water temperature setpoints. However, the move from 25°C to 27°C (77°F to 80.6°F) can have an impact on the IT equipment's power dissipation. Most IT manufacturers start to increase air-moving device speed around 25°C (77°F) to improve the cooling of the components and thereby offset the increased ambient air temperature. Therefore, care should be taken before operating at the higher inlet conditions. The concern that increasing the IT inlet air temperatures might have a significant effect on reliability is not well founded. An increase in inlet temperature does not necessarily mean an increase in component temperatures. Consider the graph in Figure G.2 showing a typical component temperature relative to an increasing ambient temperature for an IT system with constant-speed fans.

In Figure G.2, the component temperature is 21.5°C above the inlet temperature of 17°C; it is 23.8°C above an inlet ambient temperature of 38°C. The component temperature tracks the air inlet ambient temperature very closely.

Now consider the response of a typical component in a system with variable-speed fan control, as depicted in Figure G.3. Variable-speed fans decrease the fan

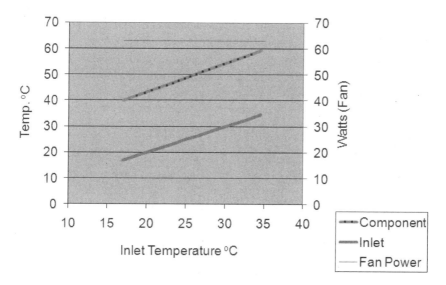

Figure G.2 Inlet and component temperatures with fixed fan speed.

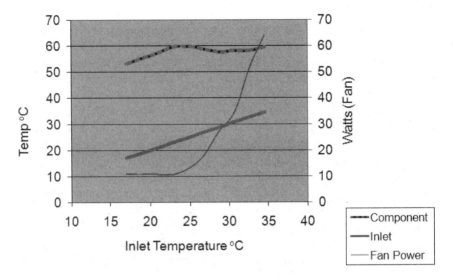

Figure G.3 Inlet and component temperatures with variable fan speed.

flow rate at lower temperatures to save energy. Ideal fan control optimizes the reduction in fan power to the point that component temperatures are still within vendor temperature specifications (i.e., the fans are slowed to the point that the component temperature is constant over a wide range of inlet air temperatures).

This particular system has a constant fan flow up to approximately 23°C. Below this inlet air temperature, the component temperature tracks closely to the ambient air temperature. Above this inlet temperature, the fan adjusts flow rate such that the component temperature is maintained at a relatively constant temperature.

This data brings up several important observations:

- Below a certain inlet temperature (23°C in the case described above), IT systems that employ variable-speed air-moving devices have constant fan power, and their component temperatures track fairly closely to ambient temperature changes. Systems that don't employ variable-speed air-moving devices track ambient air temperatures over the full range of allowable ambient temperatures.

- Above a certain inlet temperature (23°C in the case described above), the speed of the air-moving device increases to maintain fairly constant component temperatures and, in this case, inlet temperature changes have little to no effect on component temperatures and, thereby, no affect on reliability, since component temperatures are not affected by ambient temperature changes.

- The introduction of IT equipment that employs variable-speed air-moving devices has
  - minimized the effect on component reliability as a result of changes in ambient temperatures and
  - allowed for potential of large increases in energy savings, especially in facilities that deploy economizers.

As shown in Figure G.3, the IT fan power can increase dramatically as it ramps up speed to counter the increased inlet ambient temperature. The graph shows a typical power increase that results in the near-constant component temperature. In this case, the fan power increased from 11 watts at ~23°C inlet temperature to over 60 watts at 35°C inlet temperature. The inefficiency in the power supply results in an even larger system power increase. The total room power (facilities + IT) may actually increase at warmer temperatures. IT manufacturers should be consulted when considering system ambient temperatures approaching the upper recommended ASHRAE temperature specification. See Patterson (2008) for a technical evaluation of the effect of increased environmental temperature, where it was shown that an increase in temperature can actually increase energy use in a standard data center but reduce it in a data center with economizers in the cooling system.

Because of the derating of the maximum allowable temperature with altitude for Classes 1 and 2, the recommended maximum temperature is derated by 1°C/300 m (1.8°F/984 ft) above 1800 m (5906 ft).

MOISTURE LIMITS

High End

Based on extensive reliability testing of printed circuit board laminate materials, it was shown that conductive anodic filament (CAF) growth is strongly related to RH (Sauter 2001). As humidity increases, time to failure rapidly decreases. Extended periods of RH exceeding 60% can result in failures, especially given the reduced conductor-to-conductor spacings common in many designs today. The CAF mechanism involves electrolytic migration after a path is created. Path formation could be due to a breakdown of inner laminate bonds driven by moisture, which supports the electrolytic migration and explains why moisture is so key to CAF formation. The upper moisture region is also important for disk and tape drives. In disk drives, there are head flyability and corrosion issues at high humidity. In tape drives, high humidity can increase frictional characteristics of tape and increase head wear and head corrosion. High RH, in combination with common atmospheric contaminants, is required for atmospheric corrosion. The humidity forms monolayers of water on surfaces, thereby providing the electrolyte for the corrosion process. Sixty percent RH is associated with adequate monolayer buildup for monolayers to start taking on fluid-like properties. Combined with humidity levels exceeding the critical equilibrium humidity of a contaminant's saturated salt, hygroscopic corrosion product is formed, further enhancing the buildup of acid-electrolyte surface wetness and greatly accelerating the corrosion process. Although disk drives do contain internal means to control and neutralize pollutants, maintaining humidity levels below the critical humidity levels of multiple monolayer formation retards initiation of the corrosion process.

A maximum recommended dew point of 15°C (59°F) is specified to provide an adequate guard band between the recommended and allowable envelopes.

Low End

The motivation for lowering the moisture limit is to allow a greater number of hours per year where humidification (and its associated energy use) is not required. The previous recommended lower limit was 40% RH. This correlates on the psychrometric chart to 20°C (68°F) dry-bulb temperature and a 5.5°C (41.9°F) dew point (lower left) and a 25°C (77°F) dry-bulb and a 10.5°C (50.9°F) dew point (lower right). The dryer the air, the greater the risk of electrostatic discharge (ESD). The main concern with decreased humidity is that the intensity of static electricity discharges increases. These higher-voltage discharges tend to have a more severe impact on the operation of electronic devices, causing error conditions requiring service calls and, in some cases, physical damage. Static charges of thousands of volts can build up on surfaces in very dry environments. When a discharge path is offered, such as a maintenance activity, the electric shock of this magnitude can damage sensitive electronics. If the humidity level is reduced too far, static dissipative materials can lose their ability to dissipate charge and then become insulators.

The mechanism of the static discharge and the impact of moisture in the air are not widely understood. Montoya (2002) demonstrates, through a parametric study, that ESD charge voltage level is a function of dew point or absolute humidity in the air and not a function of RH. Simonic (1982) studied ESD events across various temperature and moisture conditions over a period of a year and found significant increases in the number of events (20×) depending on the level of moisture content (winter vs. summer months). It was not clear whether the important parameter was absolute humidity or RH.

Blinde and Lavioe (1981) studied electrostatic charge decay (vs. discharge) of several materials and showed that it is not sufficient to specify environmental ESD protection in terms of absolute humidity; nor is a RH specification sufficient, since temperature affects ESD parameters other than atmospheric moisture content.

The 2004 recommended range includes a dew-point temperature as low as 5.5°C (41.9°F). Discussions with IT equipment manufacturers indicated that there have been no known reported ESD issues within the 2004 recommended environmental limits. In addition, the referenced information on ESD mechanisms (Montoya 2002; Simonic 1982; Blinde and Lavio 1981) does not suggest a direct RH correlation with ESD charge creation or discharge, but Montoya (2002) does demonstrate a strong correlation of dew point to charge creation, and a lower humidity limit, based upon a minimum dew point (rather than minimum RH), is proposed. Therefore, the 2008 recommended lower limit is a line from 18°C (64.4°F) dry-bulb temperature and 5.5°C (41.9°F) dew-point temperature to 27°C (80.6°F) dry-bulb temperature and a 5.5°C (41.9°F) dew-point temperature. Over this range of dry-bulb temperatures and a 5.5°C (41.9°F) dew point, the RH varies from approximately 25% to 45%.

Another practical benefit of this change is that process changes in data centers and their HVAC systems, in this area of the psychrometric chart, are generally sensible only (i.e., horizontal on the psychrometric chart). Having a limit of RH greatly complicates the control and operation of the cooling systems and could require added humidification operation at a cost of increased energy in order to maintain an RH when the space is already above the needed dew-point temperature. To avoid these complications, the hours of economizer operation available using the 2004 guidelines were often restricted.

ASHRAE is developing a research project to investigate moisture levels and ESD with the hope of driving the recommended range to a lower moisture level in the future. ESD and low moisture levels can result in drying out of lubricants, which can adversely affect some components. Possible examples include motors, disk drives, and tape drives. While manufacturers indicated acceptance of the environmental extensions documented here, some expressed concerns about further extensions. Another concern for tape drives at low moisture content is the increased tendency to collect debris on the tape and around the head and tape transport mechanism due to static buildup.

ACOUSTICAL NOISE LEVELS

The ASHRAE 2008 recommendation to expand the environmental envelope for datacom facilities may have an effect on acoustical noise levels. Noise levels in high-end data centers have steadily increased over the years and are becoming a serious concern for data center managers and owners. For background and discussion on this, see Chapter 9, "Acoustical Noise Emissions," in ASHRAE's *Design Considerations for Datacom Equipment Centers* (2005). The increase in noise levels is the obvious result of the significant increase in cooling requirements of new, high-end datacom equipment. The increase in concern results from noise levels in data centers approaching or exceeding regulatory workplace noise limits, such as those imposed by OSHA in the U.S. or by EC directives in Europe. Empirical fan laws generally predict that the sound power level of an air-moving device increases with the fifth power of rotational speed. This means that a 20% increase in speed (e.g., 3000 to 3600 rpm) equates to a 4 dB increase in noise level. While it is not possible to predict a priori the effect on noise levels of a potential 2°C (3.6°F) increase in data center temperatures, it is not unreasonable to expect to see increases in the range of 3–5 dB. Data center managers and owners should, therefore, weigh the trade-offs between the potential energy efficiencies with the recommended new operating environment and the potential increases in noise levels.

With regard to the regulatory workplace noise limits, and to protect employees against potential hearing damage, data center managers should check whether potential changes in noise levels in their environments will cause them to trip various "action level" thresholds defined in local, state, or national codes. The actual regulations should be consulted, because they are complex and beyond the scope of this document to explain fully. For instance, when levels exceed 85 dB(A), hearing conservation programs are mandated, which can be quite costly and generally involve baseline audiometric testing, noise level monitoring or dosimetry, noise hazard signage, and education and training. When levels exceed 87 dB(A) (in Europe) or 90 dB(A) (in the U.S.), further action, such as mandatory hearing protection, rotation of employees, or engineering controls must be taken. Data center managers should consult with acoustical or industrial hygiene experts to determine whether a noise exposure problem will result from increasing ambient temperatures to the 2008 upper recommended limit.

Data Center Operation Scenarios for ASHRAE's 2008 Recommended Environmental Limits

The recommended ASHRAE guideline is meant to give guidance to IT data center operators on the inlet air conditions to the IT equipment for the most reliable operation. Four possible scenarios where data center operators may elect to operate at conditions that lie outside the recommended environmental window are listed as follows.

1. Scenario #1: Expand economizer use for longer periods of the year where hardware fails are not tolerated.

    For short periods of time it is acceptable to operate outside this recommended envelope and approach the allowable extremes. All manufacturers perform tests to verify that the hardware functions at the allowable limits. For example, if during the summer months it is desirable to operate for longer periods of time using an economizer rather than turning on the chillers, this should be acceptable, as long as the period of warmer inlet air temperatures to the datacom equipment does not exceed several days each year; otherwise, the long-term reliability of the equipment could be affected. Operation near the upper end of the allowable range may result in temperature warnings from the IT equipment.

2. Scenario #2: Expand economizer use for longer periods of the year where limited hardware fails are tolerated.

    All manufacturers perform tests to verify that the hardware functions at the allowable limits. For example, if during the summer months it is desirable to operate for longer periods of time using the economizer rather than turning on the chillers, and if the data center operation is such that periodic hardware fails are acceptable, then operating for extended periods of time near or at the allowable limits may be acceptable. This, of course, is a business decision of where to operate within the allowable and recommended envelopes and for what periods of time. Operation near the upper end of the allowable range may result in temperature warnings from the IT equipment.

3. Scenario #3: Failure of cooling system or servicing cooling equipment.

    If the system was designed to perform within the recommended environmental limits, it should be acceptable to operate outside the recommended envelope and approach the extremes of the allowable envelope during the failure. All manufacturers perform tests to verify that the hardware functions at the allowable limits. For example, if a modular CRAC unit fails in the data center, and the temperatures of the inlet air of the nearby racks increase beyond the recommended limits but are still within the allowable limits, this is acceptable for short periods of time until the failed component is repaired. As long as the repairs are completed within normal industry times for these types of failures, this operation should be acceptable. Operation near the upper end of the allowable range may result in temperature warnings from the IT equipment.

4. Scenario #4: Addition of new servers that push the environment beyond the recommended envelope.

    For short periods of time, it should be acceptable to operate outside the recommended envelope and approach the extremes of the allowable envelope. All manufacturers perform tests to verify that the hardware functions at the allowable limits. For example, if additional servers are added to the data center in an area that would increase the inlet air temperatures to the server racks above the recommended limits but adhere to the allowable limits, this should be acceptable for short periods of time until the ventilation can be improved. The length of time operating outside the recommended envelope is somewhat arbitrary, but several days would be acceptable. Operation near the upper end of the allowable range may result in temperature warnings from the IT equipment.

REFERENCES

ASHRAE. 2004. *Thermal Guidelines for Data Processing Environments*. Atlanta: American Society of Heating, Refrigerating and Air-Conditioning Engineers, Inc.

ASHRAE. 2005. *Design Considerations for Datacom Equipment Centers*. Atlanta: American Society of Heating, Refrigerating and Air-Conditioning Engineers, Inc.

Blinde, D., and L. Lavoie. 1981. Quantitative effects of relative and absolute humidity on ESD generation/suppression. *Proceedings of EOS/ESD Symposium,* vol. EOS-3, pp. 9–13.

Montoya. 2002. Sematech electrostatic discharge impact and control workshop, Austin Texas. http://ismi.sematech.org/meetings/archives/other/20021014/montoya.pdf.

Patterson, M.K. 2008. The effect of data center temperature on energy efficiency. *Proceedings of Itherm Conference, Orlando, Florida.*

Sauter, K. 2001. Electrochemical migration testing results—Evaluating printed circuit board design, manufacturing process and laminate material impacts on CAF resistance. *Proceedings of IPC Printed Circuits Expo, Anaheim, CA.*

Simonic, R. 1982. ESD event rates for metallic covered floor standing information processing machines. *Proceedings of the IEEE EMC Symposium, Santa Clara, CA,* pp. 191–98.

Telcordia. 2001. GR-3028-CORE, Thermal management in telecommunications central offices. *Telcordia Technologies Generic Requirements*, Issue 1. Piscataway, NJ: Telcordia Technologies, Inc.

Index

power delivery network 112, 115, 189

power factor correction 13, 101, 120, 129, 189

power management 13, 15, 110, 122–25, 129

power supply 3, 7, 97, 110, 113, 119–21, 123, 189, 191

power-conditioning equipment 110, 113–14, 126

present value 154–58, 188–89

pressure 11–12, 25, 27, 30, 33–34, 36, 49, 60, 62, 75–76, 78–80, 89–90, 92, 133–36, 138, 140, 142, 148–49, 181, 186, 188, 190, 192, 196, 205, 209–11, 218, 223

pressurization 59, 88, 181, 205, 210–11

processors 15, 110–11, 123–24

PSU 113–14, 118, 120–24, 126–28, 189

psychrometric 19, 56, 91, 189, 210

PUE 7–8, 160, 161, 189

PWB 112, 116–18, 189

R

rack 3, 13, 74–78, 81–82, 103, 108, 110–11, 113–14, 121, 123–24, 131–33, 142–44, 146, 182, 188–90, 203

raised floor 76, 82–83, 89–90, 158, 187, 189, 192

RCI 74–75, 83, 190, 203

redundancy 13, 88, 90, 99, 108, 122, 124, 129, 132, 159, 190

refrigerant condenser 50, 190

refrigerants 39, 133–35, 137, 140, 142, 148, 190, 208

relative humidity 12, 18–19, 23–24, 26, 35, 39, 57, 59– 60, 74, 81, 89, 91, 94–95, 181, 186–87, 189–91, 203, 211, 214, 217

reliability 1, 5, 68, 86, 92–94, 98–99, 123–24, 165, 190, 194, 196–98, 200, 204–05

relief air 56, 181, 209, 211

return air 12, 26, 46, 54, 56–57, 59–60, 63, 77, 82–84, 87, 89, 91, 181, 185, 190, 209–11

RHI 75, 83, 190–91

S

semiconductor devices 111

sensible heat 33, 89, 186, 190–91, 209

sensible heat ratio 33, 190

server 2, 5–6, 8, 78, 84, 110, 112–13, 123, 126, 132, 140, 143–45, 154, 156–57, 159, 188, 191

SHI 75, 83, 190–91